DATE DUE

			PRINTED IN U.S.A.

—NTC's—
ENCYCLOPEDIA
of
INTERNATIONAL
WEIGHTS &
MEASURES

——NTC's——
ENCYCLOPEDIA
of
INTERNATIONAL
WEIGHTS &
MEASURES

WILLIAM D. JOHNSTONE

Printed on recyclable paper

NTC Publishing Group

Lincolnwood, Illinois USA

**Cataloging in Publication Data Available from
the United States Library of Congress**

Contents

Introduction   xi

PART ONE
**UNITS OF
LENGTH**

Linear measure   2
Solar or stellar measure   18
Mariner's measure   19
Surveyor's measure   21
Cloth measure   23
Printer's measure   30
Linear units of ancient Arabia   34
Linear units of ancient Greece   35
Linear units of ancient India   40
Linear units of ancient Persia   41
Linear units of ancient Rome   42
Linear units of Japan   45
Linear units of Russia   47
Linear units of Spain   48
Linear units of Sweden   50
Linear units of Switzerland   52
Additional ancient and foreign linear units   53

PART TWO	*Square measure*	59
SURFACE	*Comparative areas of square measure units*	65
UNITS	*Surface units of ancient Rome*	67
	Surface units of Denmark	69
	Surface units of Iceland	70
	Surface units of Japan	71
	Surface units of Spain	72
	Surface units of Yugoslavia	74
	Additional ancient and foreign	
	surface units	75
	Angular measure	77
PART THREE	*Cubic measure*	82
UNITS OF	*Liquid measure*	89
CAPACITY	*Dry measure*	94
AND VOLUME	*Apothecaries' fluid measure*	101
	Cooking measure	106
	Spirits measure	110
	Comparative capacities and volumes of	
	cubic, liquid, dry, apothecaries' fluid,	
	cooking, and spirits measures	123
	Intermeasure comparison of capacity	
	and volume units	126
	Capacity and volume units of	
	ancient Arabia	129
	Capacity and volume units of	
	ancient Greece	131
	Capacity and volume units of	
	ancient India	135
	Capacity and volume units of	
	ancient Israel	136

Contents

Capacity and volume units of	
ancient Rome	137
Capacity and volume units of Argentina	140
Capacity and volume units of Austria	141
Capacity and volume units of Brazil	143
Capacity and volume units of Cyprus	144
Capacity and volume units of Denmark	145
Capacity and volume units of Egypt	146
Capacity and volume units of France	148
Capacity and volume units of Iceland	150
Capacity and volume units of Japan	151
Capacity and volume units of Libya	151
Capacity and volume units of Mexico	152
Capacity and volume units of the	
Netherlands	153
Capacity and volume units of Poland	155
Capacity and volume units of Portugal	155
Capacity and volume units of Russia	157
Capacity and volume units of Spain	160
Capacity and volume units of Sweden	163
Capacity and volume units of Switzerland	165
Additional ancient and foreign capacity	
and volume units	166

PART FOUR	*Troy weight*	170
UNITS OF	*Avoirdupois weight*	173
WEIGHT AND	*Apothecaries' weight*	186
MASS	*Intermeasure comparison of weight*	
	and mass units	187
	Weight and mass units of ancient Greece	188

Weight and mass units of ancient India   189
Weight and mass units of ancient Rome   189
Weight and mass units of Denmark   191
Weight and mass units of Egypt   192
Weight and mass units of Greece   194
Weight and mass units of Japan   195
Weight and mass units of the Netherlands   196
Weight and mass units of Poland   197
Weight and mass units of Portugal   198
Weight and mass units of Russia   199
Weight and mass units of Spain   200
Weight and mass units of Sweden   202
Weight and mass units of Yugoslavia   203
*Additional ancient and foreign weight
    and mass units*   204

PART FIVE   *Unit prefixes*   208
THE METRIC   *Metric length units*   208
SYSTEM   *Comparison of metric and customary
        length units*   211
Metric surface units   212
*Comparison of metric and customary
    surface units*   214
Metric capacity and volume units   214
*Comparison of metric and customary
    capacity and volume units*   217
Metric weight and mass units   218
*Comparison of metric mass units and
    customary weight and mass units*   219

PART SIX	*Units of count*	222
DIVERSE	*Paper measure*	226
UNITS	*Units of book size*	227
	Time measure	228
	Sound measure	241
	Units used in music	243
	Units of poetic verse	246
	Light measure	250
PART SEVEN	*Pressure units*	256
OTHER	*Temperature units*	258
DIVERSE	*Energy units*	261
UNITS	*Force units*	265
	Radiation units	267
	Viscosity units	268
	Units expressing flow of water	268
	Speed units	270
	Miscellaneous units	271
	Units of various capacities and volumes per various masses and weights	273
	Units of various lengths per various capacities and volumes	273
	Units of various masses and weights per various areas	274
	Units of various capacities and volumes per various areas	274
	Miscellaneous units relating to temperature	275
PART EIGHT	*Electrical power units*	277
ELECTRICAL	*Electrical cycle units*	280
UNITS	*Electrical frequency units*	281
	Electromotive force units	281

x *Contents*

Electrical intensity units 282
Electrical resistance units 284
Electrical conductance units 286
Electrical quantity units 286
Electrical capacity units 287
Electrical inductance units 288
Electrical units of magnetic flux 289
Electrical units of magnetic flux density 289
Magnetomotive force units 290
Units of electrical power per various areas 290
*Units involving the volt per
 various lengths* 291
*Units involving the ampere-turn
 per various lengths* 291
*Units involving the ampere per
 various areas* 292
*Units involving the coulomb per
 various areas* 292
*Units involving the ohm per
 various volumes* 293
*Units involving the mho per
 various volumes* 293
Computer Units 294

Index of units 295

General Index 316

Introduction

Measurement, that highly complicated but necessary reality, is not without its problems. This book attempts to explain these problems and facilitate their solution. It offers a condensed education and a ready reference to people in all walks of life.

Everyone uses measurement:

> Everyone who cooks, sews, drives, or shops (and who may some day have to adjust to buying carpeting by the meter, meat by the gram, and milk and gasoline by the liter)
>
> Everyone who runs a business, orders supplies, pays employees, or keeps an appointment
>
> Everyone who is studying any scientific discipline and most other disciplines as well
>
> Everyone in every profession, business, service, industry, and science—doctor, banker, surveyor, photographer, athlete, bartender, painter, teacher, builder, druggist, engineer, pilot

Let us first explore some of the problems of our present system, a system we take for granted and give little attention to in our daily routine.

Are you a boating enthusiast? And have you not at some time had to convert *statute miles* into *nautical miles*? Did you remember the conversion factor? And didn't it seem rather impractical to have one mile for land and another for sea? Are you a land developer? And have you been concerned with and confused by the relations among an *arpent,* an *acre-foot,* and a *Gunter's link*? Do you do the cooking? Surely you have at some time been forced to consider how many *drops* are in a *coffee spoon* or how many *dry pints* make up a *peck*. The list of frustrating encounters in this measuring mess is endless. My favorite example is the *mile,* known variously as the air mile, sea mile, knot, admiralty mile, nautical mile, and geographic mile, each equaling 1.853 kilometers. The Irish mile (2.05 kilometers), the Scottish mile (1.81 kilometers), the international nautical mile (1.85204 kilometers), the postal mile (7.59 kilometers), the international geographic mile (7.42 kilometers), and the meridian mile or French nautical mile (1.85201 kilometers) add to the confusion. And don't forget the statute mile or land mile (1.61 kilometers), itself known by many other names in various lands.

The *perch, pound, mil, ton, foot, quarter, year, month, ounce, line, point, octave,* and *load* are similarly complicated, as, to a lesser degree, are many other units. Some units, such as the *catty* and the *vara,* vary considerably with different locations; others, like the *bag,* vary according to how they are used. Even in antiquity, units such as the *cubit* denoted a multitude of values during their development.

The existence of British counterparts of the American units makes the use of measurement still trickier. Even the metric system, the supposed end result of measurement revision, is beset by flaws, inadequacies, and inaccuracies. But despite them, the metric system is a vast improvement over the cumbersome menagerie indicated above.

This book, then, was written in an attempt to impose some order on the great maze of units—to assist the user in measurement accuracy; to save the user time and, most likely, money; to help the user better visualize how long, how much, how heavy, or how big a particular unit is; and to simplify use of the more workable metric system.

Since the advent of civilization, and probably since the dawn of mankind, man has through necessity devised various measures to assist him in his craftsmanship and commerce. Today, virtually everything he does, creates, trades, duplicates, transports, or stores involves measurement.

In the beginning, man used his own body as a measuring standard to advance his primitive technology, communication, defense, and bartering. The length of his arm became a unit of length now called the *yard*; the length of his foot, obviously enough, became the *foot*; the size of his hand gave rise to the *palm, hand,* and *span*; and the amount of land his oxen could plow in one day led to the *acre*. In ancient Greece, the *stadium* became a length unit probably geared to the length of a foot race. And the weight of a carob seed was the basis of the *carat*.

Problems arose, however, because men's arms, feet, and hands varied in size; foot races were of different lengths; one ox was more efficient than another; and one carob seed's weight varied from the next.

Such inaccurate units necessarily gave way to more precise measures; some units were cast aside and new ones added until whole unwieldy systems were in common use. Today, we are inundated with units, most of which serve little or no useful purpose. A further word on accuracy involves the Uncertainty Principle, created by Werner Heisenberg, (1901—1976), a German physicist, who stated that there is a definite limit to man's ability to measure. For example, when you measure the temperature of a liquid, the immersion of the thermometer changes the temperature; and when you measure an electric current, some of the current flows through the meter.

The most practical system, known as the *international system*, was first proposed in 1670 by Gabriel Mouton, Vicar of St. Paul in Lyons, France, but was not adopted by that country until 1795. The system was probably perfected from the decimal or "ten" system that reached Europe from the Orient in the fourteenth century.

In the United States, Congress passed the Surveyor's Act in 1799, the first federal weights and measures law. By 1832, it had adopted the *Troy pound* as the standard for coinage and the *Winchester bushel* as the basis for customs duties.

July 27, 1866 brought about the Kassen Act, making metric weights and measures "lawful throughout the United States of America" for legal and commercial transactions, its use being voluntary. Great Britain had passed a similar act in 1864. By 1875, the United States and sixteen other nations signed the Metric Convention, or Treaty of the Meter. The National Bureau of Standards was established in the United States in 1901.

A modernized version of the *metric system* was developed in 1960 during the international General Conference on Weights and Measures. This new system is also called *Le Système international d'unités,* or the *International System of Units,* abbreviated SI. It is still most commonly referred to as the metric system.

By 1975, the U.S. Congress had passed the Metric Conversion Act to foster voluntary conversion. In 1988, it amended and strengthened the act with the new Omnibus Trade and Competiveness Act. Then, three years later, Metric Usage in Federal Government Programs, was enacted by Executive Order; and in 1992, the U.S. Fair Packaging & Labeling Act was passed.

Although the process of conversion and public acceptance of conversion to the metric system is slow, significant changes have occurred. Several departments of the U.S. government, for the most part, have switched, including the military and the space program. The optical, chemical, pharmaceutical, engineering, electrical, computer, photographic, and tobacco industries have gone metric.

Most research and development is conducted in metric units. Manufacturers of automobiles, tractors, and earth moving equipment, including their suppliers, switched to metrics in the 1970s. The liquor industry reduced the number of its container sizes from 53 to 7 after converting to metrics. Today, in fact, almost one half of all U.S. products are designed, built, and sold using metric specifications. Nevertheless, the United States still cannot call itself a metric country.

Today, three systems are in general use: the *metric system,* the *United States customary system,* and the *British imperial system.*

The *metric system, (Le Système International d'Unités),* employs the physical fundamental quantities of the meter, kilogram, second, ampere, kelvin, and candela, and the geometric supplementary units of the radian and steradian. This system is augmented by many metric derived units, such as the liter, hertz, newton, pascal, joule, watt, and lux; as well as the lumen, coulomb, volt, ohm, tesla, weber, siemens, farad, henry, and others.

The *United States customary system* is the most familiar and widely used system in the United States and, with its fundamental units of the yard and the avoirdupois pound, is the source of the current measurement mess. Only the countries of Barbados, Gambia, Ghana, Liberia, Oman, Sierra Leone, Southern Yemen, Tonga, and Trinidad presently use this old system (together with their own units based on local folklore).

The *British imperial system,* now becoming outdated in Britain, also incorporates the fundamental units of the yard and pound, but with slightly different values from those of the United States. In the British system, from which the American system evolved, the dry capacity units and liquid units are the same; in the American system they differ. (For example, the *liquid pint* and *dry pint* in Britain each equals 0.568 cubic decimeter, about 12 percent larger than the U.S. liquid pint of 0.473 cubic decimeter and slightly larger than the U.S. dry pint of 0.551 cubic decimeter. And the common U.S. *short ton* equals 9.072 quintals, but the British *long ton* equals 10.161 quintals. Also, the U.S. *inch* equals 2.540005 centimeters, while the British inch equals 2.539998 centimeters. Is this important to the man in the street? Often not; but it can be. An American bolt doesn't necessarily fit a British nut, or a British light bulb an American socket.) In 1965, Britain, as a condition for becoming a member of the European Common Market, began a transition to the metric system in its trade and commerce.

With all systems of measure, certain basic definitions apply. *Fundamental units* are quantities whose scale of measurement is arbitrarily assigned and is independent of the scales of other quantities; all other quantities are *derived units,* derived from the original fundamental units. For example, the foot is a derived unit, as it is 1/3 yard (the fundamental unit). A *unit of measurement* is a precisely defined quantity in terms of which the magnitudes of all other quantities of the same kind can be stated. A *standard of measurement* is an object which, under specified conditions, defines, represents, or records the magnitude of a unit.

It was Richard I of England in the twelfth century who passed the first law requiring actual standards for length and capacity.

These standards were made of iron and were kept by sheriffs and magistrates. Today, the standard of the meter, for instance, is a bar of 90 percent platinum and 10 percent iridium; virtually every country keeps a copy. The international prototype meter No. 27 is the twenty-seventh copy of the original and is kept by the International Bureau of Weights and Measures. Kilogram No. 20 is a cylinder about 1.5 inches high and is the standard of mass. This standard is so accurate that when, in 1937, at age 50, it was compared with the original, it was found to differ only by 1/50,000,000.

The metric system itself is simple and precise. It is in conversion from customary units that complications and confusion arise. Adequate motivation for such a shift exists, mainly in trade and technology. And yet the goals of the United States are lofty, at best—to use the metric system for *all* products, processes, and information in its dealings with international commerce; to educate the young in the measurement system of the future, and to change the habits of more than 280 million Americans who've been quite content with the old system for more than 200 years. The goal of the U.S. Federal Highway Administration for instance, is to include both metric and customary units (kilometers and miles) on highway speed and distance signs by September 30, 1996—a huge and costly undertaking.

But when the time comes that the metric system is in full use, by all peoples the world over, some customary and ancient units will undoubtedly persist. Surely a *first down* in American football will remain 10 yards, and large parcels of land will still be thought of in terms of *acres*. Similarly, certain time-honored phrases will never change: It's hard to imagine "give him a centimeter, and he'll take a kilometer" or "a gram of prevention is worth a kilogram of cure." And there will be those who cling to the old ways. Reprinted in a U.S. magazine some years ago was this objection from an Australian: "No doubt many readers have noticed as I have that since eggs went metric, they have been pale in yolk colour and lacking freshness. This clearly shows that chooks [chickens] cannot adjust to laying different size eggs. We tamper with nature at our peril."

Nevertheless, metrics may soon be here to stay, and we must adjust to it. And in the future a new system of time and angular measure may also be proposed and adopted and have to be adjusted to. About 4,000 years ago, the Babylonians decided that the convenient way to divide up a circle was into 360 parts. In one respect, it was convenient, as 360 can be divided evenly by 2, 3, 4, 5, 6, 8, 9, 10, 12, 15, 18, 20, 24, 30, 36, 40, 45, 60, 72, 90, 120, and 180. This 360 system evolved into the 360 degrees of angular measure, the geographer's longitudinal and latitudinal divisions of a spherical earth, the second-minute-hour system of the clock, and, remotely, the calendar.

Decimalizing could instantly ease many kinds of calculation. The circle could be divided into 100 parts with decimal subdivisions. In timekeeping, a metric system called *decimal time* has long been under consideration, with the day divided into 20 hours of 100 minutes each, and 100 seconds to the minute. The section Time Measure expands on present and future time and calendar systems.

Also with the metric system comes a switch to the Celsius temperature scale, known more familiarly as the centigrade scale, from the common Fahrenheit system. Water would freeze at 0 instead of 32 degrees, boil at 100 instead of 212 degrees; body temperature would be normal at 37 instead of 98.6 degrees; and a 25-degree day would be comfortable. The U.S. National Weather Service may then report wind velocities in kilometers per hour.

Before describing the organization and uses of this book, additional clarification of a few terms is important. Such terms are not necessarily pertinent to any individual unit, but contribute to understanding the overall concept of measurement.

The *net* in connection with a weight means the weight of the substance alone—and only net values are dealt with in this book; *gross* weight includes the weight of the substance's container. Gross weight is also described as the weight of something without deduction for tare, tret, or waste.

A *heaped measure* indicates the substance is piled high in its container, while *struck measure* is measured to the top level of the container. All applicable units in this book are struck.

In considering pressure and weight, a *dead load* in, say, a building, is what the floor itself weighs plus the building weight it holds. A *live load* includes the furniture, machinery, people, and the force of moving bodies upon that floor.

Many people use the terms *weight* and *mass,* and *capacity* and *volume,* interchangeably, but there are differences. Weight, for instance, is actually the force exerted on a body by gravity. That is why objects which have mass are said to be "weightless" in space, when there are no effects of gravity. Further explanations of the differences are clarified under their respective sections.

Any unit is generally one of four types:

Decimal, divisible into tenths, a system handed down by the Chinese and Egyptians, and what the metric system is really all about

Duodecimal, divisible into twelfths, a method used by the Romans

Binary, divisible into halves, then quarters, eighths, and so on, system of the Hindus

Sexagesimal, divisible into sixtieths, a system used by the Babylonians and reflected today in the division of our clocks, calendars, and circles

To indicate the accuracy of the material values given each unit, the following classification codes are offered:

(e) indicates the number is exact

(r) indicates the number is exact as far as it goes, but is rounded off to the nearest five, six, or seven places

(a) indicates the number is an approximation of its real value and is used in those cases where adequate information was difficult to obtain

A word should also be said about spelling; the me*ter* or the original French and British met*re*; the dra*chm* or dra*m,* the li*ter* or lit*re*? For the most part, this book uses the common American versions and does not attempt to include the many international spelling variations of the many units. There are, of course, exceptions.

The book is written primarily for the United States resident and offers some help to the American overseas traveler. It might also be useful, however, in enabling peoples of other lands to better understand the entire picture of measurement and in particular those measurement complexities peculiar to the United States. In the sections concerning foreign countries, I have included all units I have found to be in use today, according to direct, reliably communicated, or published observation. Those countries or regions absent from one or more parts of the book, or only miscellaneously represented, may be assumed to employ standard metric units virtually exclusively.

All computations are believed to be completely accurate, computed from the stated quantities of units found in numerous published sources. In some cases, however, adequate information was unavailable, and the unit's value was computed in relation to other known units. The reader will find a great many instances where alternative values are cited for particular units in the supplementary information below their listings. The vagaries of regional and national customs, the complications inherent in usage in different professions and over many centuries, and the contradictions in published works of ostensibly equal authority make it almost impossible in a compendium as broad and comprehensive as this one to define the bases of such variances. A great many seem to be merely the result of rounding off fractional values, but in other cases differences are so drastic as to imply a radical shift in usage or a contrary interpretation. Rather than delineate these differences, I have simply given in each listing the value that seems to me the most proper, accurate, or prevalent and followed it with whatever alternate values have also come to my attention. In many instances, the geographical names given with the units reflect the historical period in which the terms were valid.

William D. Johnstone

NTC's
ENCYCLOPEDIA
of
INTERNATIONAL
WEIGHTS &
MEASURES

PART ONE

Units of length

Length units are presented in Part One:

Linear measure

Solar or stellar measure

Mariner's measure

Surveyor's measure

Cloth Measure

Printer's measure

Units of ancient Arabia, Greece, India, Persia, and Rome

Units of Japan, Russia, Spain, Sweden, Switzerland, and other foreign units, both current and ancient

Length, for the purpose of these sections, may be thought of as the shortest straight-line distance between two points, and a *unit of length* as the entire or an increment of a particular distance. According to Webster, length is the longest dimension of any object, a specific extent or distance in that dimension.

The earliest known unit of length, its name lost in antiquity, was probably a unit used by the megalithic tomb builders in Britain about 2300 B.C. that equaled about 2.72±0.003 feet.

Linear measure

The following section shall be referred to as containing linear units, but also commonly known is the title of "long" measure.

1 MICRO-ANGSTROM

0.000001 (e) angstrom

The microangstrom is so small that almost 32 trillion make up ⅛ inch.

1 FERMI

0.010000 (e) X-unit

0.000010 (e) angstrom

10^{-12} (e) millimeter

The fermi, named for Enrico Fermi (1901–1954), American physicist, is used in subatomic measure and equals 10^{-13} centimeter. The term fermi also occurs under Square Measure in Part Two and under Metric Length Units in Part Five.

1 X-UNIT, SIEGBOHN UNIT, MILLIANGSTROM

100.00000 (e) fermis

0.001000 (e) angstrom

10^{-10} (e) millimeter

This unit was named for Georg Siegbohn, physicist. The terms also occur under Metric Length Units in Part Five.

1 ANGSTROM, ANGSTROM UNIT

1,000,000.00000 (e) microangstroms

100,000.00000 (e) fermis

1,000.00000 (e) X-units

0.100000 (e) millimicron

0.000100 (e) micron

0.0000001 (e) millimeter

The unit was named for A. J. Angström (1814–1874), physicist, and is used in the measure of waves of light and electricity. The term angstrom also occurs under Metric Length Units in Part Five, and under Light Measure in Part Six.

1 MILLIMICRON, MILLICRON	10.00000 (e) angstroms
	0.039370 (r) microinch
	0.001000 (e) micron
	0.0000010 (e) millimeter

The unit also occurs under Metric Length Units in Part Five.

1 MICROINCH 10^{-6} (e) inch
0.0000254 (r) millimeter

1 MICRON   10,000.00000 (e) angstroms
1,000.00000 (e) millimicrons
39.37008 (r) microinches
0.0010000 (e) millimeter

The term micron also occurs under Metric Length Units in Part Five.

1 SILVERSMITH'S POINT   0.000250 (e) inch
0.0063500 (r) millimeter

1 MIL   25.40005 (e) microns
0.072271 (r) point
0.023301 (r) agate
0.012000 (e) line
0.001000 (e) inch
0.0254001 (r) millimeter

The mil is similar to the artillery mil, a unit of angle used to measure large-bore guns and equal to the angle subtended by an arc of 1/6400 circumference; the artillery mil has replaced the old infantry mil, an angle subtended by an arc of 1/1000 circumference.

Other units similar in length or name include the landmil, which in Denmark equals 4.68 (a) statute miles and 4,000 (e) favn. In Sweden a mil equals 10.0 (e) kilometers and is known as a Swedish mile, but in old Sweden the mil equaled 6.64 statute miles. The mila a landi in Iceland equals 4.68 (a) statute miles and 24,000 (e) fet, and the meile in Prussia equaled 4.68 (a) statute miles and 24,000 (e) fuss.

The mil is used in measuring wire and the bores of firearms. The term mil also occurs under Angular Measure in Part Two.

1 POINT

351.46049 (r) microns
13.83700 (e) mils
0.193720 (r) agate
0.165878 (r) line
0.083333 (r) pica
0.013837 (r) inch
0.3514605 (r) millimeter

The point is also considered to equal 0.013830 (e) inch and 1/10 line; in percentages, a point equals 1 percent or 1/100 part.

The point began as a French measure, when the common pica was divided into twelve parts or points. The system was adopted by the U.S. Type Founders Association in 1886.

The term point also occurs under Printer's Measure in Part One and under Angular Measure in Part Two.

1 OUNCE

0.015625 (e) inch
0.3968758 (r) millimeter

The ounce is used to measure the thickness of shoe leather, but for the sole of the shoe, the iron unit is used. The term ounce also occurs under Troy Weight, Avoirdupois Weight, and Apothecaries' Weight in Part Four, and under Spirits Measure in Part Three.

1 HAIRS-
BREADTH,
HAIRBREADTH

0.020833 (r) inch
0.5291582 (r) millimeter

1 AGATE

1,814.21203 (r) microns
71.42553 (r) mils
5.16214 (r) points
0.857170 (r) line
0.071428 (r) inch
1.8142857 (r) millimeters

The agate, called a ruby in Britain, is used as the standard height of type in printing and is considered also to equal 0.9186 inch.

1 LINE

83.33333 (r) mils

6.02250 (r) points

1.16667 (r) agates

0.501875 (r) pica

0.083333 (r) inch

0.006944 (r) foot

2.1166667 (r) millimeters

The line is known as a gramme in Greece. A royal gramme in Greece, however, equals 1 (e) millimeter. A line in France is known as a ligne and equals 1/12 pouc, 0.0888 (r) inch, and 2.256 (r) millimeters.

Other units similar in length or name include the ligne or linie, which in Switzerland equals 0.8202 (r) inch and 2.0833 (r) millimeters. A ligne or liniya in Russia equals 0.1 (e) inch. A linie in Austria equals 1/144 fuss and 0.087 inch, and a linie in Bavaria equals 1/144 fuss and 0.080 inch. A lina in Iceland equals 1/144 fet, 0.0858 (r) inch, and 2.18 (r) centimeters.

Still other similar units are the linea, which in Chile equals 1/432 vara and 0.076 (r) inch. A linea in Argentina and Paraguay equals 1/432 vara and 0.0789 (r) inch; a linea in Mexico equals 1/432 vara and 0.0764 (r) inch; and a linea in Spain equals 1/432 vara and 0.0762 (r) inch.

A linha in Portugal equals 1/144 pe and 0.0902 (r) inch. A linja in Poland equals 1/144 stopa, 0.0787 (r) inch, and 2 (e) millimeters.

The linje in Denmark equals 1/144 fod and 0.0858 (r) inch, and a linje in Sweden equals 1/144 fot and 0.117 (r) inch. The niu in Siam equals 1/12 keup, 0.82 (r) inch, and 2.083 (r) centimeters. The keup in Siam equals 9.84 (r) inches, but in old Siam equaled 12 (e) niu and 10 (e) inches.

The term line is also used in Cloth Measure and Printer's Measure, Part One, and in Electrical Units of Magnetic Flux, Part Eight.

1 BARLEYCORN, SIZE

333.33333 (r) mils

24.09000 (r) points

1 Barleycorn,	4.00000 (e) lines
size (continued)	0.333333 (r) inch
	0.111111 (r) palm
	0.083333 (r) hand
	0.027778 (r) foot `
	0.018519 (r) cubit
	8.4666667 (r) millimeters

The unit is of ancient origin and began from the size of a grain of barley. It is the basis for the measure of footwear. The term barleycorn also occurs under Troy Weight in Part Four.

1 DIGIT   0.750000 (e) inch

1.9050038 (r) centimeters

The digit is a British unit derived from the breadth of a man's finger, but may have roots in the ancient Hebrew unit, the ezba, which equaled about 0.73 inches.

1 INCH, PRIME   25,400.05000 (e) microns

4,000.00000 (e) silversmith's points

1,000.00000 (e) mils

72.27000 (r) points

64.00000 (e) ounces

48.00000 (e) hairsbreadths

23.24635 (r) agates

12.00000 (e) lines

3.00000 (e) barleycorns

1.33333 (r) digits

0.333333 (r) palm

0.250000 (e) hand

0.083333 (r) foot

0.055556 (r) cubit

0.027778 (r) yard

0.005051 (r) rod

2.5400050 (e) centimeters

Originally the inch was the width of a man's thumb. In the fourteenth century, Edward II of England decreed: "The length of an inch shall be equal to three grains of barley, dry and round, placed end to end lengthwise."

Presently, the inch has been corrected to equal 41,929.399 wavelengths of krypton 86, measured at 760 millimeters pressure and 15 degrees centigrade.

The inch described above is also known as the old standard inch, international inch, and Canadian inch, which equaled 2.54 (e) centimeters. The British inch or imperial inch equals 2.539998 (r) centimeters.

The inch is known as a liin or toll in Estonia, a duim or duime in Russia, a daktylos in Greece, a djuim in the Byelorussian Soviet Socialist Republic, a pulgada in the Philippines, and a tuuma in Finland.

A duim equals 1 (e) centimeter in the Netherlands. A royal daktylos equals 1 (e) centimeter in Greece, while a daktylos in ancient Greece equaled 1/24 cubit 0.76 (r) inch, and 1.93 (r) centimeters. A pulgada in Mexico equals 1/12 pie and 0.916 (r) inch, and a pulgada in Spain equals 1/12 pie and 0.914 (r) inch.

The Cape inch of South Africa equals 1.033 inches, while the tomme of Denmark equals 1.03 (r) inches, 1/12 fod, and 26.15 (r) centimeters. The thumlunger of Iceland equals 1.03 (r) inches and 1/12 fet. The zoll of West Germany equals 1.03 (r) inches, but the zoll of old Prussia equaled 1/12 fuss and 1.03 (r) inches, and the zoll of Switzerland equals 1.181 (r) inches and 3.0 (e) centimeters.

The mulvelyk of Hungary equals 1.04 (r) inches and 2.63 (r) centimeters. The pollegada of Portugal and Brazil equals 1.08 (r) inches and 2.75 (r) centimeters. The pouce of old France equaled 1.066 (r) inches and 27.07 (r) centimeters, and the pouce of Switzerland equals 3.0 (e) centimeters.

The term inch is also used in Cloth Measure and Printer's Measure, Part One.

1 HANDS-BREADTH, HANDBREADTH

The unit is considered to equal 2.5 to 4.0 inches and probably had the same origins as, and has been used interchangeably with, the linear units hand and palm.

1 PALM

9.00000 (e) barleycorns

3.00000 (e) inches

0.750000 (e) hand

0.250000 (e) foot

0.083333 (r) yard

7.6200150 (e) centimeters

The palm, an obsolete unit, was originally derived from the length of a man's palm, from the heel of the hand to the tip of the middle finger, a length longer than the modified palm defined above. It has also been considered merely another version of the linear units hand and handsbreadth.

The palm is known as a roupi in Cyprus when used in textiles and as a moot in Calcutta. A moot also equals 1/12 guz. The term palm also occurs under Metric Length Units in Part Five.

1 HAND

12.00000 (e) barleycorns

4.00000 (e) inches

1.33333 (r) palms

0.333333 (r) foot

0.222222 (r) cubit

0.111111 (r) yard

1.0160020 (e) decimeters

The hand is an ancient unit, originally determined by the width, little finger to thumb, of a man's hand. It is now used for measuring the height of a horse at the withers and for measuring bananas, a small bunch being a hand and a large bunch a stem.

1 SHAFTMENT,
SHATHMONT

6.00000 (e) inches

1.5240030 (e) decimeters

This is an ancient Scottish unit and was originally measured from the tip of a man's extended thumb across the breadth of the palm.

1 FOOT

12,000.00000 (e) mils

144.00000 (e) lines

36.00000 (e) barleycorns

12.00000 (e) inches

4.00000 (e) palms

3.00000 (e) hands

0.666667 (r) cubit

0.333333 (r) yard

0.060606 (r) rod

0.001515 (r) furlong

3.0480060 (e) decimeters

The foot was originally derived from the length of a man's foot, and the term is also used in Mariner's Measure and Surveyor's Measure, Part One.

The foot is also known as a survey foot, used in mapping. In Canada the foot is called a pied; in Argentina, a pie; in Haiti, a pied anglais; in El Salvador, a tercia; and in Russia, a foute or fut, which also equals 1/7 sagene.

A pie in Paraguay equals ⅓ vara, 11.36 (r) inches, and 28.85 (r) centimeters; a pie in Spain equals 10.97 (r) inches and 27.86 (r) centimeters; a pie in Mexico equals 12.0 (e) pulgadas and 10.992 (r) inches.

A pied in Switzerland equals 11.81 (r) inches and 30.0 (e) centimeters; a pied in Belgium, a varying unit, usually equals 12.79 inches and 32.49 centimeters; and a pied de roi of old France equaled 12.0 (e) pouces, 1.066 (r) feet, and 32.48 (r) centimeters.

The pe of Portugal and Brazil equals 1.08 feet; the shaku of Japan equals 1/6 ken and 0.994 (r) feet; and the voet of the Netherlands equals 11.14 (r) inches.

A French foot equals 12.8 inches, 1.067 (r) feet, and 3.2512064 (e) decimeters. A fod in Denmark equals ½ alen, 1.030 (r) feet, and 31.38 (r) centimeters; a fet in Iceland equals 12.36 inches, 31.39 (r) centimeters, and 37.67 (r) centimeters; a fot in Sweden equals 1/10 stang, 11.69 inches, and 29.69 (r) centimeters; and a fot in Norway equals 12.35 inches, 31.37 (r) centimeters, and 37.643 (r) centimeters.

The fuss in Austria equals 1.037 (r) feet and 31.61 (r) centimeters; a fuss in Bavaria equals 11.5 inches and 29.19 (r) centimeters; a fuss or schun in Switzerland equals 11.81 inches; and a fuss or Rhine foot of old Prussia equaled 1.030 feet and 31.38 (r) centimeters.

The stopa in Bohemia equals 11.65 inches; the stopa in Poland equals 11.34 inches; and the stopa in Yugoslavia equals 1/6 khvat and 12.44 inches.

The old standard foot equaled 3.048 (e) decimeters, while a foot equaled 12.5 inches in ancient Greece and 12.4 inches in old Prussia. The ancient and obsolete Roman foot equaled 11.62 inches or 2.9514858 (r) decimeters, 11.64 inches, and 11.65 inches, and was also known as a pes or peda. The foot of ancient Arabia, Assyria, and Persia equaled 12.6 inches.

1 CUBIT

54.00000 (e) barleycorns

18.00000 (e) inches

6.00000 (e) palms

---

1 Cubit *(continued)*	4.00000 (e) hands 4.5720090 (e) decimeters

The cubit is an ancient and obsolete unit derived from the distance between the elbow and the tip of the middle finger. The cubit is also considered to have equaled 18.24 inches. The Bible cubit was 21.8 inches. In Assyria a cubit equaled 21.6 inches; in Egypt, 20.6 inches; in Israel, 17.6 inches; in Greece, 18.3 inches; in Rome, 17.5 inches. It was given many additional values during its development.

The hasta of Singapore and the hath or moolum of India equal 18.0 inches, while the asta of Malacca equals 17.99 inches. The hasta of ancient India equaled 25.26 inches, and the cubito of Somaliland equals 1/7 top, 22 inches, and 56 centimeters.

1 BITE

A bite of bricks is ten bricks stacked face to face, standing on end in a row 25 inches long for carrying purposes.

1 VARA

33.33333 (r) inches

2.77778 (r) feet

0.925926 (r) yard

8.4666667 (r) decimeters

In Spanish and Portuguese, the word *vara* means "forked pole," or prop, staff, or wand.

The vara described above is used in California, Texas, and Louisiana. But in Colombia it equals 31.5 inches or 80 (e) centimeters; in Honduras, 32.87 (r) inches; in Spain, 32.88 (r) inches and 32.913 (r) inches; in Chile, Peru, Guatemala, and El Salvador, 32.913 (r) inches; in Cuba, 33.39 (r) inches and 33.386 (r) inches; in Uruguay, 33.63 (r) inches; in Brazil, 43.31 (r) inches and 43.308 (r) inches; in Argentina, 34.094 (r) inches; and in Paraguay, 34.08 (r) inches.

1 YARD

108.00000 (e) barleycorns

36.00000 (e) inches

12.00000 (e) palms

9.00000 (e) hands

3.00000 (e) feet

2.00000 (e) cubits

0.181818 (r) rod
0.004545 (r) furlong
9.1440180 (e) decimeters

There are three popular explanations for the origin of the yard: (1) in northern Europe a yard equaled the length of the girdle the Anglo-Saxons wore; (2) in southern Europe, a yard equaled the length of a double cubit; and (3) in the twelfth century, Henry I of England fixed the yard as the distance between his nose and the thumb of his outstretched arm.

The standard for the yard described above, also known as a British yard or imperial yard, is 3,600,000/3,937,008 meters, as established in 1959, and replaces the old standard yard of 3,600,000/3,937,014 meters.

The yard is known as a war in Aden and a wari in Kenya. A yard in Mexico equals 83.8 (r) centimeters. A yard in India is known as a guz and also equals 16 (e) geerah or gireh and 144 jaob or jow. The guz of Calcutta also equals ½ danda; while the guz of Persia equaled 4.0 (e) charac and 40.96 (r) inches.

The term yard is also used in Mariner's Measure, Surveyor's Measure, and Cloth Measure, Part One, and occurs under Spirits Measure in Part Three.

1 PACE, STEP

The pace for common measuring purposes equals 3.0 or 3.3 feet. In the United States military, a quick-time pace is 30.0 inches, and double time is 36.0 inches. An itinerary pace equals 5.0 feet, and a Roman pace or geometric pace equaled 58.1 inches and 5.0 Roman feet. The term pace is also used in Square Measure, Part Two.

1 REED

108.00000 (e) inches

6.00000 (e) cubits

27.4320540 (e) decimeters

The reed is an ancient and obsolete Hebrew unit. It is also considered to have equaled 8.8 feet and was also known as a qahen.

1 ROD, POLE, PERCH

198.00000 (e) inches

16.50000 (e) feet

5.50000 (e) yards

0.025000 (e) furlong

5.0292099 (e) meters

During the Middle Ages, the length of a rod in Britain was determined by lining up 16 men and measuring the combined length of all their left feet.

A perch in Quebec equals 17.8 (r) feet or 6.39 (r) yards. The perche of Belgium equals 7.11 (r) yards and 6.50 meters, and the perche of Switzerland equals 3.0 (e) meters. A rod or chang in China equals 11.75 feet, and a chang in Mongolia equals 10.5 feet.

The terms rod, pole, and perch also occur under Surveyor's Measure in Part One and under Cubic Measure in Part Three.

1 ROPE

240.00000 (e) inches
20.00000 (e) feet
6.66667 (r) yards
0.083333 (r) wrap
0.055556 (r) lea
6.0960120 (e) meters

The rope, an old British unit, was probably another form of the rod in the meaure of land. It varied in length from 5.5 to 7.0 yards.

1 WRAP

2,880.00000 (e) inches
240.00000 (e) feet
80.00000 (e) yards
12.00000 (e) ropes
0.666667 (r) lea
7.3152144 (e) decameters

The wrap is also considered to equal 3,000 (e) yards.

1 LEA

360.00000 (e) feet
18.00000 (e) ropes
1.50000 (e) wraps
1.0972822 (r) hectometers

The lea is also considered to equal a league, a linear length of 3.0 miles. The term lea also occurs under Cloth Measure in Part One.

1 STADIUM

625.00000 (e) Roman feet
605.20867 (r) feet

403.47244 (r) cubits
201.73622 (r) yards
0.125000 (e) Roman mile
1.8446786 (r) hectometers

The stadium, possibly also spelled stadian, probably originated with the length of a foot race in ancient Rome, and is also considered to have equaled 202 yards.

1 FURLONG

660.00000 (e) feet
220.00000 (e) yards
40.00000 (e) rods
0.125000 (e) statute mile
0.041667 (r) league
2.0116840 (r) hectometers

The furlong originally was the length of a plowed furrow of an ordinary field. The term furlong is also used in Surveyor's Measure, Part One.

1 MILE,
STATUTE MILE,
LAND MILE

5,280.00000 (e) feet
3,520.00000 (e) cubits
1,760.00000 (e) yards
320.00000 (e) rods
8.00000 (e) furlongs
0.868978 (r) meridian mile
0.868976 (r) international nautical mile
0.868383 (r) geographic mile
0.333333 (r) league
0.216919 (r) international geographic mile
1.6093472 (r) kilometers

The mile in England equaled 5,000 feet until the year 1500 when it was changed to 5,280 feet to make surveying easier. This way the furlong, the commonest land measure of the time, could be divided into a mile eight times. About 1575, Queen Elizabeth I established this new value by law.

The legal mile in Argentina is the milla. The mile in ancient Arabia equaled 7,093 feet. The term statute mile is also used in Surveyor's Measure, Part One.

1 SCOTTISH   5,940.00000 (e) feet
MILE   1,980.00000 (e) yards
                   1,926.48857 (r) Scottish ells
                   360.00000 (e) rods
                   9.00000 (e) furlongs
                   1.12500 (e) statute miles
                   0.883929 (r) Irish mile
                   0.325643 (r) international geographic mile
                   1.8105156 (r) kilometers

1 MERIDIAN   6,076.10333 (r) feet
MILE,   2,025.36778 (r) yards
FRENCH   1.15078 (r) statute miles
NAUTICAL MILE 0.999999⁺ (r) international nautical mile
                   0.999315 (r) geographic mile
                   0.249622 (r) international geographic mile
                   0.016667 (r) degree
                   1.8520046 (r) kilometers

The meridian mile was established by international agreement in 1954 and closely approximates 1/60 degree of the earth's meridian. Many similar units should be noted.

In Finland, the meripeninkulma equals 1.15 statute miles. In old Turkey, the berri equaled 1.04 statute miles. In Finland, the sjomil equals 1.151 statute miles and 1,852 meters, while the sjomil of Iceland equals 6,085.95 (r) feet and 1,855 meters. In Sumatra, the paal equals 1.151 statute miles and 1,852 meters, while in Java, the paal equals 1,648 yards.

In Argentina, the milla equals 1.0 (e) statute mile; in Venezuela, 1.154 statute miles; in Nicaragua, 1.159 statute miles; and in Honduras, 1.149 statute miles. The mila of Poland equals 5.30 statute miles, and the mijl of the Netherlands equals 0.621 (r) statute mile and 1.0 (e) kilometer.

The miglio of Italy equals 1.0 (e) kilometer; the miglio of old Rome, however, equaled 0.925 statute mile; and the miglio of old Naples equaled 1.38 statute miles. The milha of Portugal equals 1.28 statute miles and 2,066 meters, and the milha of Brazil equals 1.21 statute miles and 1,955 meters.

The meile of Austria equals 24,000 fuss and 4.71 statute miles; the meile of Prussia equals 24,000 fuss and 4.68 statute miles; and the meile or merfold of

Hungary equals 5.19 statute miles. The unit is probably the same as the mille marin or French geographic mile.

1 INTERNA-
TIONAL
NAUTICAL MILE,
INTERNATIONAL
AIR MILE

6,076.11549 (r) feet

2,025.37183 (r) yards

1.15078 (r) statute miles

1.00002 (r) meridian miles

0.999317 (r) geographic mile

0.249627 (r) international geographic mile

0.016667 (r) degree

1.8520416 (r) kilometers

The international nautical mile is considered the most accurate mile measure to date using the earth's meridian.

1 GEOGRAPHIC
MILE,
NAUTICAL MILE,
KNOT, AIR MILE,
SEA MILE

6,080.27000 (e) feet

2,026.75667 (r) yards

368.50121 (r) rods

9.21253 (r) furlongs

1.15157 (r) statute miles

1.00067 (r) meridian miles

1.00066 (r) international nautical miles

0.249797 (r) international geographic mile

0.016681 (r) degree

1.8532699 (r) kilometers

The geographic mile originally was designed to equal 1 minute or 1/21600 of a great circle of the earth and has survived as another version of the meridian mile. In 1929, the International Hydrographic Bureau proposed a length of 6,076.097 (e) feet to succeed both the meridian and geographic miles, which in turn have been replaced by the international nautical mile.
   The geographic mile is also known as an admiralty mile in Britain. All the unit's names also occur under Mariner's Measure in Part One; the term knot is also used in Speed Units, Part Seven.

1 IRISH MILE

6,720.00000 (e) feet

2,240.00000 (e) yards

407.27273 (r) rods

1 *Irish mile*	10.18182 (r) furlongs
(continued)	1.27273 (r) statute miles
	1.13131 (r) Scottish miles
	0.313583 (r) international geographic mile
	2.0482600 (r) kilometers

1 LEAGUE, LEA,	15,840.00000 (e) feet
LAND LEAGUE,	5,280.00000 (e) yards
CANNON-SHOT	960.00000 (e) rods
DISTANCE	24.00000 (e) furlongs
	3.00000 (e) statute miles
	0.868383 (r) marine league
	0.650757 (r) international geographic mile
	0.044175 (r) degree
	4.8280415 (r) kilometers

The term lea is defined earlier in this section. The term cannon-shot distance was derived from the distance a cannon shot could be fired in the late eighteenth century and gave rise to the 3-mile offshore territorial limit the United States once employed in international law, and which now is 12 miles. The league is an ancient unit, used by the Gauls, that has varied from 2.5 to 4.5 statute miles.

Another unit similar in distance and name is the legua, which in Uruguay equals 3.2 (r) statute miles; in Cuba, 2.635 statute miles; in Argentina, 3.23 statute miles; in Mexico, 2.60 statute miles; in Paraguay, 2.69 statute miles; in Chile, 5,400 (e) varas and 2.81 statute miles; and in Spain, 6,666.667 varas and 3.46 statute miles.

The legoa in Brazil equals 20,000 (e) pes and 4.10 statute miles, and the legoa in Portugal equals 24 (e) estadios and 3.85 statute miles. The lieue of old France equaled 2,280.3 toises and 2.76 statute miles. The kos or koss of India varies from 1.5 to 3.0 statute miles. The farsakn of Arabia equals 3.001 statute miles, and the farsakn of Persia equaled 3.88 statute miles.

1 INTERNA-	24,340.80000 (e) feet
TIONAL	8,113.60000 (e) yards
GEOGRAPHIC	1,475.20000 (e) rods
MILE	368.80000 (e) furlongs

4.61000 (e) statute miles

4.00599 (r) meridian miles, international nautical miles

4.00324 (r) geographic miles

1.53667 (r) leagues

0.066767 (r) degree

7.4190906 (r) kilometers

The international geographic mile equals 4.60 statute miles in Indonesia, and was designed to equal 1/15 degree of the earth's meridian, which it approximates.

The landmil of Denmark equals 4.68 statute miles; the mila a landi of Iceland equals 24,000 (e) fet and 4.68 statute miles; and the meile of Prussia equaled 24,000 (e) fuss and 4.68 statute miles. The chausseemeile of Germany equals 4.6 statute miles.

1 POSTAL MILE

4.71400 (a) statute miles

7.5864627 (r) kilometers

The postal mile is an Austrian unit; the meile of Austria, possibly a postal mile rounded off, equals 4.710 statute miles. A mile in Russia equals 4.64 statute miles; in Denmark, 4.68 statute miles; in Sweden, 6.642 statute miles; in Norway, 7.019 statute miles. In Finland, a peninkulma equals 6.21 statute miles.

1 DEGREE

364,566.92940 (r) feet

69.04677 (r) statute miles

60.00120 (r) meridian miles

60.00000 (r) international nautical miles

59.95900 (r) geographic miles

23.01559 (r) leagues

19.98633 (r) marine leagues

14.97767 (r) international geographic miles

111.1224972 (r) kilometers

The degree is also considered to equal 69.169 statute miles. The degree is equal to 1/60 of the earth's circumference, but a degree of latitude equals 68.708 statute miles at the equator and 69.403 statute miles at the poles, as determined by the International Astronomical Union Ellipsoid of 1964, and a degree of longitude equals 69.171 statute miles and at the poles is zero.

The term degree also occurs under Angular Measure in Part Two and under Electrical Cycle Units in Part Eight.

Solar or stellar measure

Solar or stellar measures include those units used by astrophysicists and astronomers to measure the vast reaches of outer space.

1 LIGHT-SECOND

186,282.39600 (r) statute miles

0.200399 (r) astronomical unit

299,793.0524119 (r) kilometers

The light-second is the distance light travels in 1 second. The value of 186,282.42 ±.06 statute miles was used prior to November 1972, when it was corrected to the value listed above.

**1 ASTRONOMI-
CAL UNIT**

92,955,750.00000 (a) statute miles

499.00445 (r) light-seconds

0.000016 (r) light-year

0.000005 (r) parsec

149,598,072.9864000 (a) kilometers

The astronomical unit is defined as the average distance from the earth to its sun.

1 LIGHT-YEAR

5,878,627,117,143.65784 (r) statute miles

31,557,609.54000 (r) light-seconds

63,241.13481 (r) astronomical units

0.306843 (r) parsec

9,460,753,090,819.2177426 (r) kilometers

9.4607531 (r) terameters

The light-year is defined as the distance light travels in 1 sidereal year, representing here an exact year of 31,557,609.54 seconds, times the value for the speed of light (186,282.3960 statute miles per second or 299,792.5 kilometers per second, as originally determined by the International Astronomical Union in

Hamburg in 1964, adopted in 1968, and corrected in November 1972 to 186,282.3960 statute miles per second ± 3.6 feet per second).

1 PARSEC

19,173,488,480,298.00000 (a) statute miles
206,264.70000 (a) astronomical units
3.26160 (a) light years
30,856.78000 (a) gigameters

The parsec is also considered to equal 3.259 (a) light years. The name is a combination of the first syllables of *parallax* and *second*, and the unit is defined as the distance of an imaginary star from the earth to form a maximum angle or parallax of 1 second or 1/3600 degree.

The length of a kiloparsec (1,000 parsecs) is so great that, as an example, it would take an airplane traveling at a constant speed of 600 statute miles per hour more than 3.6479 million years to cover the distance.

Mariner's measure

This section is devoted to measurements used for marine craft at sea and for air travel.

1 FOOT

0.333333 (r) yard
0.166667 (r) fathom
0.001389 (r) cable length
3.0480060 (e) decimeters

The term foot is also used in Linear Measure and Surveyor's Measure, Part One.

1 YARD

3.00000 (e) feet
0.500000 (e) fathom
0.004167 (r) cable length
9.1440180 (e) decimeters

The term yard also occurs under Linear Measure, Surveyor's Measure, and Cloth Measure in Part One, and under Spirits Measure in Part Three.

1 FATHOM
6.00000 (e) feet
2.00000 (e) yards
0.008333 (r) cable length
18.2880360 (e) decimeters

The fathom is used in measuring marine depth and was derived from the width of a man's outstretched arms, fingertip to fingertip, which was originally established as being 79.20 inches. The name comes from the Danish faedn, which means outstretched arms. In Japan, a ken equals 1.988 yards.

1 MARK TWAIN
The Mark Twain equals 2.0 (e) fathoms, or the minimum safe clearance for the steam wheelers and other large river boats of the late 1800s.

1 CABLE LENGTH
720.00000 (e) feet
240.00000 (e) yards
120.00000 (e) fathoms
0.118416 (r) nautical mile
2.1945643 (e) hectometers

The cable length defined above is common with the U.S. Navy. The ordinary cable length, or cable, equals 600 (e) feet. The British cable length equals 608 (e) feet, 607.61033 (r) feet, and 1/10 French nautical mile, and was derived from the usual length of a ship's cable in the days of sailing ships. Depth of water now is measured on large ships by sound waves.

1 NAUTICAL MILE, KNOT, GEOGRAPHIC MILE, AIR MILE, SEA MILE
6,080.27000 (e) feet
2,026.75667 (r) yards
1,013.37833 (r) fathoms
8.44482 (r) cable lengths
0.333333 (r) marine league
1.8532699 (r) kilometers

The nautical mile is also known as the admiralty mile in Britain. All the above terms are also used in Linear

Measure, Part One, and the term knot occurs under Speed Units in Part Seven.

The name knot originated as a division in a special line called a ship's log line. This line was tied to a small log that kept it afloat. At intervals, the line was knotted in the same proportion that 28 seconds are to an hour, or about 47 feet apart. The number of knots that ran off a reel of line in 28 seconds was the number of nautical miles the ship was traveling in an hour.

Similar units include the sjomil of Finland, which equals 1.151 statute miles, and the sjomil or sea mile of Iceland, which equals 6,085.95 (r) feet and 1,852 meters. The ri of Japan equals 36 (e) cho and 2.44 statute miles; the marine ri of Japan equals 1.15 statute miles; and the obsolete kenning equaled 20.21 statute miles and 32.2 to 33.8 kilometers.

1 MARINE LEAGUE

18,240.81000 (e) feet

6,080.27000 (e) yards

3,040.13500 (e) fathoms

25.33446 (r) cable lengths

3.00000 (e) nautical miles

0.050044 (r) degree

5.5598097 (r) kilometers

Surveyor's measure

Surveyor's measure is used by surveyors, engineers, land developers, and assessors in the surveying, development, and appraising of land.

1 LINK, GUNTER'S LINK, SURVEYOR'S LINK

7.92000 (e) inches

0.660000 (e) foot

0.220000 (e) yard

0.040000 (e) rod

2.0116840 (e) decimeters

The link is attributed to the English mathematician Edmund Gunter (1581–1626), inventor of the sector and scale.

1 FOOT,
SURVEY FOOT

12.00000 (e) inches
1.64141 (r) links
0.333333 (r) yard
0.060606 (r) rod
0.015152 (r) chain
0.001515 (r) furlong
3.0480060 (e) decimeters

The foot, for purposes of surveying, also equals the engineer's link and Ramden's link. The term survey foot occurs also under Linear Measure in Part One, and the term foot is also used in Linear Measure and in Mariner's Measure, Part One.

1 YARD

36.00000 (e) inches
4.92424 (r) links
3.00000 (e) feet
0.181818 (r) rod
0.045455 (r) chain
0.004545 (r) furlong
9.1440180 (e) decimeters

The term yard also occurs under Linear Measure, Mariner's Measure, and Cloth Measure in Part One, and under Spirits Measure in Part Three.

1 ROD, POLE,
PERCH

198.00000 (e) inches
25.00000 (e) links
16.50000 (e) feet
5.50000 (e) yards
0.250000 (e) chain
0.025000 (e) furlong
5.0292099 (e) meters

The terms rod, pole, and perch are used in Linear Measure, Part One; the term perch also occurs under Square Measure in Part Two and under Cubic Measure in Part Three.

1 CHAIN,	100.00000 (e) links
GUNTER'S	66.00000 (e) feet
CHAIN,	22.00000 (e) yards
SURVEYOR'S	4.00000 (e) rods
CHAIN	0.100000 (e) furlong
	0.012500 (e) statute mile
	2.0116840 (r) decameters

The engineer's chain or Ramden's chain also equals 100 links, but each link equals 1.0 foot.

1 FURLONG	1,000.00000 (e) links
	660.00000 (e) feet
	220.00000 (e) yards
	40.00000 (e) rods
	10.00000 (e) chains
	0.125000 (e) statute mile
	2.0116840 (r) hectometers

The furlong was derived from the length of an average plowed furrow, and the unit is also used in horse racing. The term furlong is also used in Linear Measure, Part One.

1 STATUTE MILE	8,000.00000 (e) links
	5,280.00000 (e) feet
	1,760.00000 (e) yards
	320.00000 (e) rods
	80.00000 (e) chains
	8.00000 (e) furlongs
	1.6093472 (e) kilometers

The term statute mile is also used in Linear Measure, Part One, and is known by several other names listed in that section.

Cloth measure

Cloth measures are those units pertinent to the manufacture and marketing of cloth, fabric, and virtually any wearing apparel or linen.

1 IRON
0.020833 (r) inch
0.5291592 (r) millimeter

The iron is used in the measure of the leather in the soles of shoes; the ounce is used to measure all other leather in shoes. The terms iron and ounce occur under Linear Measure in Part One.

1 LINE
0.025000 (e) inch
0.6350013 (e) millimeter

The line is used in the measure of buttons. The term occurs also under Linear Measure and Printer's Measure in Part One and under Electrical Units of Magnetic Flux in Part Eight.

1 INCH
48.00000 (e) irons
40.00000 (e) lines
0.444444 (r) nail
0.222222 (r) finger
0.111111 (r) span
0.027778 (r) yard
0.027027 (r) Scottish ell
0.022222 (r) English ell
0.018519 (r) thread
2.5400050 (e) centimeters

The term inch is also used in Linear Measure and Printer's Measure, Part One, and is known by several other names listed in those sections.

1 NAIL
2.25000 (e) inches
0.500000 (e) finger
0.250000 (e) span
0.062500 (e) yard
0.060811 (r) Scottish ell
0.050000 (e) English ell
0.041667 (r) thread
5.7150113 (r) centimeters

The nail is known as a geerah or gireh in India, where it also equals 1/16 guz.

---

1 FINGER

4.50000 (e) inches
2.00000 (e) nails
0.500000 (e) span
0.125000 (e) yard
0.121622 (r) Scottish ell
0.100000 (e) English ell
0.083333 (r) thread
1.1430023 (r) decimeters

The finger was derived from the length of a man's middle finger. The term finger also occurs under Spirits Measure in Part Three.

1 SPAN,
QUARTER

9.00000 (e) inches
4.00000 (e) nails
2.00000 (e) fingers
0.333333 (r) Flemish ell
0.250000 (e) yard
0.243243 (r) Scottish ell
0.200000 (e) English ell
0.166667 (r) thread
2.2860045 (e) decimeters

The span was derived from the width of a spread hand. The quarter is also considered to equal ¼ statute mile and ¼ of many units. The term quarter also occurs under Dry Measure in Part Three and under Troy Weight and Avoirdupois Weight in Part Four. In Abyssinia, a sinzer also equals 9.0 (e) inches.

1 FLEMISH ELL

27.00000 (e) inches
6.00000 (e) fingers
3.00000 (e) spans
0.750000 (e) yard
6.8580135 (e) decimeters

Units similar to this ell include the elle of Estonia, which equals 21 inches; the elle of Switzerland, which equals 23.62 inches and 60.0 (e) centimeters; and the el of Surinam, which equals 2.26 feet.

1 YARD 36.00000 (e) inches
16.00000 (e) nails
8.00000 (e) fingers
4.00000 (e) spans
1.33333 (r) Flemish ells
0.972973 (r) Scottish ell
0.800000 (e) English ell
0.666667 (r) thread
0.025000 (e) bolt (cotton)
0.014286 (r) bolt (wool)
9.1440180 (e) decimeters

The term yard also occurs under Linear Measure, Mariner's Measure, and Surveyor's Measure in Part One, and under Spirits Measure in Part Three.

1 SCOTTISH ELL 37.00000 (e) inches
16.44444 (r) nails
8.22222 (r) fingers
4.11111 (r) spans
1.37037 (r) Flemish ells
1.02778 (r) yards
0.844444 (r) English ell
0.685185 (r) thread
0.025694 (r) bolt (cotton)
0.014683 (r) bolt (wool)
9.3980185 (e) decimeters

The ell, whether Scottish or from some other locality, has been used since biblical times and was probably derived by measuring from the hand, around the elbow, and back to the hand again. In the tenth century, King Edgar of England standardized the ell to equal a yard, but many variations still remain.

1 ENGLISH ELL 45.00000 (e) inches
20.00000 (e) nails
10.00000 (e) fingers
5.00000 (e) spans
1.66667 (r) Flemish ells
1.25000 (e) yards

1.21622 (r) Scottish ells
0.833333 (r) thread
0.031250 (e) bolt (cotton)
0.017857 (r) bolt (wool)
11.4300225 (e) decimeters

1 THREAD   54.00000 (e) inches
24.00000 (e) nails
12.00000 (e) fingers
6.00000 (e) spans
2.00000 (e) Flemish ells
1.50000 (e) yards
1.45946 (r) Scottish ells
1.20000 (e) English ells
0.037500 (e) bolt (cotton)
0.021428 (r) bolt (wool)
13.7160270 (e) decimeters

1 BOLT   1,440.00000 (e) inches
(COTTON)   640.00000 (e) nails
320.00000 (e) fingers
160.00000 (e) spans
53.33333 (r) Flemish ells
40.00000 (e) yards
38.91892 (r) Scottish ells
32.00000 (e) English ells
26.66667 (r) threads
0.571429 (r) bolt (wool)
0.333333 (r) lea (cotton and silk)
0.046997 (r) hank (cotton)
3.6576072 (e) decameters

A bolt of cotton is also known as a piece. A bolt when used to measure wallpaper equals 16 yards.

1 BOLT (WOOL)   2,520.00000 (e) inches
1,120.00000 (e) nails
560.00000 (e) fingers

1 Bolt (wool)	280.00000 (e) spans
(continued)	93.33333 (r) Flemish ells
	70.00000 (e) yards
	68.10811 (r) Scottish ells
	56.00000 (e) English ells
	46.66667 (r) threads
	1.75000 (e) bolts (cotton)
	0.233333 (r) lea (wool)
	0.125000 (e) hank (wool)
	6.4008126 (e) decameters

1 LEA, SKEIN	4,320.00000 (e) inches
(COTTON AND	1,920.00000 (e) nails
SILK)	960.00000 (e) fingers
	480.00000 (e) spans
	160.00000 (e) Flemish ells
	120.00000 (e) yards
	116.75676 (r) Scottish ells
	96.00000 (e) English ells
	80.00000 (e) threads
	3.00000 (e) bolts (cotton)
	0.140992 (r) hank (cotton)
	0.007937 (r) spindle (cotton)
	10.9728216 (e) decameters

The lea is also known as a cut when used to measure wool (see below). A skein is also a loosely coiled length of yarn; a large skein is a hank.

1 CUT (WOOL)	10,800.00000 (e) inches
	300.00000 (e) yards
	4.28571 (r) bolts (wool)
	0.500000 (e) heer
	0.125000 (e) hank (wool)
	2.7432054 (e) hectometers

A cut of wool, indicating yarn size, is based on the number of fixed-length hanks per avoirdupois pound, varying with the material and number count of the material. Number 1 count means that 1 meter of yarn weighs 1 gram. A 1 cut of asbestos or glass yarn has 100 yards to the pound; a 1 cut of woolen yarn has 300 yards to the pound, as above; and 1 cut of linen generally equals 30 yards per pound.

This system evolved because Francis I of France decided that a denier (from the weight of an old Roman coin) would be used to measure cloth goods and that a thread of 1-denier size would measure 4.5 million yards to a pound—about 2,530 miles long.

A denier may also be described as a unit of fineness for rayon, nylon, and silk yarns, based on a standard of 50.0 milligrams per 450 meters of yarn. A typp is a unit of count for all yarns, such as 1,000 yards per pound.

1 HANK (WOOL)
560.00000 (e) yards
8.00000 (e) bolts (wool)
1.86667 (r) cuts (wool)
0.933333 (r) heer
5.1206501 (r) hectometers

1 HEER
600.00000 (e) yards
8.57143 (r) bolts (wool)
2.00000 (e) cuts (wool)
1.07143 (r) hanks (wool)
0.041667 (r) spindle (wool)
5.4864108 (e) hectometers

The heer is a unit so old that its origin is lost in antiquity. The unit now is used in the measure of linen and wool yarn.

1 HANK (COTTON)
840.00000 (e) yards
817.29730 (r) Scottish ells
672.00000 (e) English ells
560.00000 (e) threads
21.00000 (e) bolts (cotton)
7.00000 (e) leas (cotton and silk)
0.055556 (r) spindle (cotton)
7.6809751 (r) hectometers

1 SPINDLE	14,400.00000 (e) yards
(WOOL)	205.71429 (r) bolts (wool)
	48.00000 (e) cuts (wool)
	25.71428 (r) hanks (wool)
	24.00000 (e) heers
	131.6738592 (e) hectometers

A spindle is also a rod on which yarn is wound for spinning.

1 SPINDLE	15,120.00000 (e) yards
(COTTON)	378.00000 (e) bolts (cotton)
	126.00000 (e) leas (cotton and silk)
	18.00000 (e) hanks (cotton)
	138.2575522 (r) hectometers

Printer's measure

Printing originated with the Chinese, significant contributions being made by Pi Sheng, who invented the first movable type, made of clay, in 1058–1061, and Wang Chen, who made the first wooden type in 1314. Metal type was first used in Korea about 1403. Movable metal type in Europe is attributed to Johann Gutenberg (c. 1400–1468) of Germany, who printed the Bible.

Since typesetting is based on type sizes of so many *points*, most units in this section are sizes of type, including the point unit itself. Exceptions are *agate line, milline, inch, stick, em,* and *en,* which are differently related to type size. Not included in this section, but used in the printing industry, are nameless type sizes equaling 6, 24, 30, 36, 42, and 60 points.

| **1 POINT** | 0.165878 (r) line |
| | 0.013837 (r) inch |

0.006919 (r) stick

0.3514605 (r) millimeter

The point is also considered to equal 0.013830 (e) inch and 0.013837 (e) inch. The term point also occurs under Linear Measure in Part One and under Angular Measure in Part Two.

1 EXCELSIOR 3.00000 (e) points

0.041511 (r) inch

1.0543815 (r) millimeters

1 BRILLIANT 3.50000 (e) points

0.048430 (r) inch

1.2301118 (r) millimeters

The brilliant is also considered to equal 4.0 (e) points.

1 DIAMOND 4.00000 (e) points

0.055348 (r) inch

1.4058420 (r) millimeters

The diamond is also considered to equal 4.5 (e) points.

1 PEARL 5.00000 (e) points

0.069185 (r) inch

1.7573075 (r) millimeters

1 AGATE 5.50000 (e) points

0.076104 (r) inch

1.9330328 (r) millimeters

The agate is known as a ruby in Britain. The term agate is also used in Linear Measure, Part One.

1 NONPAREIL 6.00000 (e) points

0.083022 (r) inch

2.1087680 (r) millimeters

1 LINE 6.02250 (r) points

0.083333 (r) inch

1 Line *(continued)*	0.041667 (r) stick 2.1166667 (r) millimeters

The term line is also used in Linear Measure and Cloth Measure, Part One, and in Electrical Units of Magnetic Flux, Part Eight.

An agate line equals one column of type in width by ¼ inch in height. A milline equals 1 agate line, a column wide (about 2.25 inches), appearing in 1 million copies of a publication, or more simply:

$$\frac{\text{agate lines} \times \text{number of copies}}{1,000,000}$$

1 MIGNONETTE   6.50000 (e) points
0.089941 (r) inch
2.2844933 (r) millimeters

The mignonette is known as an emerald in Britain.

1 MINION   7.00000 (e) points
0.096859 (r) inch
2.4602285 (r) millimeters

1 BREVIER   8.00000 (e) points
0.110696 (r) inch
2.8116890 (r) millimeters

1 BOURGEOIS   9.00000 (e) points
0.124533 (r) inch
3.1631495 (r) millimeters

1 LONG PRIMER   10.00000 (e) points
0.138370 (r) inch
3.5146049 (r) millimeters

1 SMALL PICA   11.00000 (e) points
0.152207 (r) inch
3.8660654 (r) millimeters

1 PICA	12.00000 (e) points
	2.00000 (e) nonpareils
	0.166044 (r) inch
	4.2175259 (r) millimeters
1 ENGLISH	14.00000 (e) points
	2.00000 (e) minions
	0.193718 (r) inch
	4.9204470 (r) millimeters
1 COLUMBIAN	16.00000 (e) points
	2.00000 (e) breviers
	0.221392 (r) inch
	5.6233680 (r) millimeters
1 GREAT PRIMER	18.00000 (e) points
	2.00000 (e) bourgeoises
	0.249066 (r) inch
	6.3262890 (r) millimeters
1 PARAGON	20.00000 (e) points
	2.00000 (e) long primers
	0.276740 (r) inch
	7.0292098 (r) millimeters
1 MERIDIAN	44.00000 (e) points
	0.608828 (r) inch
	15.4642616 (r) millimeters
1 CANON	48.00000 (e) points
	0.664176 (r) inch
	16.8701035 (r) millimeters
1 INCH, COLUMN INCH	72.27000 (r) points
	12.00000 (e) lines
	0.500000 (e) stick
	2.5400050 (e) centimeters

The inch is also considered to equal 72.0 (e) points, and is used to measure the height of a column, rather than the width. A standard height, called a type high, is 0.9186 inch.

The term inch also occurs under Linear Measure and Cloth Measure in Part One, and is known by other names and values listed in those sections.

1 STICK

144.54000 (r) points
24.00000 (e) lines
2.00000 (e) inches
5.0800100 (e) centimeters

1 EM

An em is the portion of a line of type taken up by the letter *M* in whatever type size is used. An en is half the size of an em.

Linear units of ancient Arabia

The following units may prove useful to the archeologist and to students of the arts and history.

1 ASSBAA

0.787400 (r) inch
0.250000 (e) cabda
0.065618 (r) foot
20.0000000 (e) millimeters

For practical purposes, the value of 20 (e) millimeters is given the unit, but it probably differed from this slightly and is also considered to have equaled 1/16 foot, which the millimeter value closely approximates.

1 CABDA

4.00000 (e) assbaas
3.14960 (r) inches
0.262472 (r) foot
80.0000000 (e) millimeters

1 ARABIAN FOOT

12.56640 (a) inches
1.04720 (a) feet
31.9187188 (r) centimeters

This foot of ancient Arabia, Assyria, and Persia is also considered to have equaled 12.6 inches.

1 QASAB   150.79680 (a) inches
12.56640 (a) feet
12.00000 (e) Arabian feet
38.3024626 (r) decimeters

The qasab was also used in ancient Assyria, Chaldea, and Persia.

1 GHALVA   252.00000 (a) yards
230.4292536 (a) meters

The ghalva was also used in ancient Assyria and Persia.

1 MILLE   6,283.20000 (a) feet
6,000.00000 (e) Arabian feet
2,094,40000 (a) yards
1.19000 (r) statute miles
1.9151231 (r) kilometers

The term mille also occurs under Linear Measure in Part One.

1 BARID   14.28000 (r) statute miles
12.00000 (e) milles
22.9814776 (r) kilometers

The new barid of Arabia equals 4.0 (e) farsakh and 12.0 (e) statute miles. The barid defined above is also considered to have equaled 14.32 statute miles.

Linear units of ancient Greece

The following units may prove useful to the archeologist and to students of the arts and history.

1 DAKTYLOS 0.759000 (a) inch
0.500000 (e) condylos

1 *Daktylos*	0.250000 (e) palaiste
(continued)	0.125000 (e) dichas
	0.062600 (e) pous
	19.2786380 (r) millimeters

The ancient daktylos equaled 1/24 pechya or Grecian cubit, while the modern royal daktylos equals 1.0 (e) centimeter.

1 CONDYLOS
2.00000 (e) daktyloi
1.51800 (a) inches
0.500000 (e) palaiste
0.250000 (e) dichas
0.125000 (e) pous
38.5572759 (r) millimeters

1 PALAISTE
4.00000 (e) daktyloi
3.03600 (a) inches
2.00000 (e) condyloi
0.500000 (e) dichas
0.250000 (e) pous
77.1145518 (r) millimeters

The palaiste is also considered to have equaled 3.04 inches and 77.15 millimeters.

1 DICHAS
8.00000 (e) daktyloi
6.07200 (a) inches
4.00000 (e) condyloi
2.00000 (e) palaistai
0.500000 (e) pous
0.200000 (e) bema
154.2291036 (r) millimeters

1 SPITHAME
9.10800 (a) inches
0.750000 (e) pous
23.1343655 (r) centimeters

The spithame is also considered to have equaled 7.5 inches and 19.05 centimeters.

1 POUS

16.00000 (e) daktyloi
12.14400 (a) inches
8.00000 (e) condyloi
4.00000 (e) palaistai
2.00000 (e) dichades
1.33333 (r) spithamai
1.01200 (a) feet
0.800000 (e) pygon
0.666667 (r) pechya
0.400000 (e) bema
0.166667 (r) orgyia
0.010000 (e) plethron
30.8458207 (r) centimeters

Another popular version of the ancient foot or pous is 12.5 inches.

1 PYGON

15.18000 (a) inches
1.26500 (a) feet
1.25000 (e) podes
0.833333 (r) pechya
0.500000 (e) bema
0.208333 (r) orgyia
38.5572759 (r) centimeters

1 PECHYA,
GRECIAN CUBIT

18.21600 (a) inches
1.50000 (e) podes
1.20000 (e) pyga
0.480000 (e) bema
0.250000 (e) orgyia
46.2687311 (r) centimeters

The pechya is also considered to have equaled 18.3 inches.

1 BEMA

30.36000 (a) inches
5.00000 (e) dichades
2.53000 (a) feet
2.50000 (e) podes

1 Bema	2.00000 (e) pyga
(continued)	1.66667 (r) pechyai
	0.416667 (r) orgyia
	7.7114552 (r) decimeters

1 XYLON   54.64800 (a) inches
4.55400 (a) feet
4.50000 (e) podes
3.60000 (e) pyga
3.00000 (e) pechyai
1.44000 (e) bemata
0.750000 (e) orgyia
13.8806193 (r) decimeters

The xylon is also considered to have equaled 55.0 inches.

1 ORGYIA   72.86400 (a) inches
24.00000 (e) palaistai
12.00000 (e) dichades
6.07200 (a) feet
6.00000 (e) podes
4.80000 (e) pyga
4.00000 (e) pechyai
2.40000 (e) bemata
1.33333 (r) xylona
0.060000 (e) plethron
1.8507492 (r) meters

1 ACAENA   121.44000 (a) inches
20.00000 (e) dichades
10.12000 (a) feet
10.00000 (e) podes
8.00000 (e) pyga
0.100000 (e) plethron
3.0845821 (r) meters

The acaena is also considered to have equaled 10.0 (e) feet.

1 PLETHRON   200.00000 (e) dichades
101.20000 (a) feet
100.00000 (e) podes
80.00000 (e) pyga
66.66667 (r) pechyai
16.66667 (r) orgyiai
10.00000 (e) acaenai
0.166667 (r) stadion
0.083333 (r) diaulos
30.8458207 (r) meters

1 STADION   607.20000 (a) feet
600.00000 (e) podes
100.00000 (e) orgyiai
6.00000 (e) plethra
0.500000 (e) diaulos
0.083333 (r) dolichos
185.0749242 (r) meters

The stadion of ancient Greece is also considered to have equaled 622.0 feet; in modern Greece, a stadion equals 1.0 (e) kilometer.

1 DIAULOS   1,214.40000 (a) feet
1,200.00000 (e) podes
404.80000 (a) yards
200.00000 (e) orgyiai
12.00000 (e) plethra
2.00000 (e) stadia
0.166667 (r) dolichos
370.1498486 (r) meters

1 DOLICHOS   72.00000 (e) plethra
12.00000 (e) stadia
6.00000 (e) diauloi
1.37000 (a) statute miles
2.2208991 (r) kilometers

The dolichos is also considered to have equaled 1.38 statute miles.

1 STATHMOS   116.00000 (e) dolichoi
15.89200 (a) statute miles
25.5757457 (r) kilometers

The stathmos, also used in ancient Persia, is also considered to have equaled 16.0 (e) statute miles.

Linear units of ancient India

The following units may prove useful to the archeologist and to students of the arts and history.

1 ANGULA   1.05250 (a) inches
0.083333 (r) vitasti
2.6733553 (r) centimeters

1 VITASTI   12.63000 (a) inches
12.00000 (e) angulas
32.0802632 (r) centimeters

1 HASTA, CUBIT   25.26000 (a) inches
24.00000 (e) angulas
2.00000 (e) vitastis
0.250000 (e) denda
6.4160526 (r) decimeters

The term cubit is also used in Linear Measure, Part One.

**1 DENDA,
DHANUSH**   101.04000 (a) inches
96.00000 (e) angulas
8.42000 (a) feet
8.00000 (e) vitastis
4.00000 (e) hastas
2.5664211 (r) meters

1 CROSA   8,000.00000 (e) hastas
2,000.00000 (e) dendas

3.18939 (r) statute miles
5.1328421 (r) kilometers

1 GAVYUTI   16,000.00000 (e) hastas
4,000.00000 (e) dendas
6.37879 (r) statute miles
2.00000 (e) crosas
10.2656842 (r) kilometers

Linear units of ancient Persia

The following units may prove useful to the archeologist and to students of the arts and history.

1 CHARAC   10.24000 (a) inches
0.250000 (e) guz
0.067725 (r) cane
26.0096512 (r) centimeters

1 GUZ   40.96000 (a) inches
4.00000 (e) characs
0.270899 (r) cane
104.0386048 (r) centimeters

1 CANE   151.20000 (a) inches
12.60000 (a) feet
3.69141 (r) guz
38.4048756 (r) decimeters

1 CHEBEL   828.00000 (a) inches
69.00000 (a) feet
20.21485 (r) guz
5.47619 (r) canes
21.0312414 (r) meters

1 STADIUM   704.00000 (a) feet
10.20289 (r) chebels

| 1 *Stadium* | 0.033333 (r) parasang |
| (*continued*) | 214.5796224 (r) meters |

The stadium unit also occurs under Linear Measure, Linear Units of Ancient Greece, and Linear Units of Ancient Rome in Part One.

1 PARASANG   30.00000 (e) stadia
4.00000 (a) statute miles
6.4373888 (r) kilometers

1 MANSION   80,000.00000 (e) Assyrian feet
15.91000 (a) statute miles
3.97750 (r) parasangs
25.6047140 (r) kilometers

The mansion was also used in Assyria and Chaldea. The Assyrian foot, used in ancient Assyria, Persia, and Chaldea, is considered to have equaled 8.663 feet.

Linear units of ancient Rome

The following units may prove useful to the archeologist and to students of the arts and history.

1 DIGITUS   0.750000 (e) uncia
0.727500 (a) inch
0.250000 (e) minor palmus
0.083333 (r) major palmus
1.8478537 (r) centimeters

1 UNCIA   1.33333 (r) digiti
0.970000 (a) inch
0.333333 (r) minor palmus
0.111111 (r) major palmus
0.083333 (r) pes
2.4638049 (r) centimeters

1 (MINOR) PALMUS	4.00000 (e) digiti
	3.00000 (e) unciae
	2.91000 (a) inches
	0.333333 (r) major palmus
	0.250000 (e) pes
	7.3914146 (r) centimeters

1 (MAJOR) PALMUS	12.00000 (e) digiti
	9.00000 (e) unciae
	8.73000 (a) inches
	3.00000 (e) minor palmi
	0.750000 (e) pes
	2.2174244 (r) decimeters

1 PES, ROMAN FOOT	16.00000 (e) digiti
	12.00000 (e) unciae
	11.64000 (a) inches
	4.00000 (e) minor palmi
	1.33333 (r) major palmi
	0.800000 (e) palmipes
	0.666667 (r) cubitus
	0.400000 (e) gradus
	0.200000 (e) passus
	0.100000 (e) decempeda
	29.5656584 (r) centimeters

The Roman foot is also considered to have equaled 11.62 inches, 2.9514858 decimeters, and 11.65 inches.

1 PALMITES, PALMES	14.55000 (a) inches
	1.25000 (e) pedes
	0.833333 (r) cubitus
	0.500000 (e) gradus
	36.9570730 (r) centimeters

The palmites is also considered to have equaled 39.96 (e) centimeters.

1 CUBITUS,	17.46000 (a) inches
CUBIT	1.50000 (e) pedes
	1.20000 (e) palmites
	0.600000 (e) gradus
	0.300000 (e) passus
	0.150000 (e) decempeda
	44.3484876 (r) centimeters
1 GRADUS	2.50000 (e) pedes
	2.42500 (a) feet
	2.00000 (e) palmites
	1.66667 (r) cubiti
	0.808333 (r) yard
	0.500000 (e) passus
	0.250000 (e) decempeda
	0.020833 (r) actus
	7.3914146 (r) decimeters
1 PASSUS, PACE,	5.00000 (e) pedes
ROMAN PACE	4.85000 (a) feet
	4.00000 (e) palmites
	3.33333 (r) cubiti
	2.00000 (e) gradus
	1.61667 (r) yards
	0.500000 (e) decempeda
	0.041667 (r) actus
	14.7828292 (r) decimeters
1 DECEMPEDA	10.00000 (e) pedes
	9.70000 (a) feet
	6.66667 (r) cubiti
	4.00000 (e) gradus
	3.23333 (r) yards
	2.00000 (e) passus
	0.083333 (r) actus
	2.9565658 (r) meters

1 ACTUS
120.00000 (e) pedes
48.00000 (e) gradus
38.80000 (a) yards
24.00000 (e) passus
12.00000 (e) decempedae
0.192000 (e) stadium
35.4787901 (r) meters

1 STADIUM
625.00000 (e) pedes
605.20867 (a) feet
125.00000 (e) passus
5.20833 (r) acti
0.125000 (e) milliarium
184.7853650 (r) meters

The terms stadian and stadium occur also under Linear Measure in Part One.

1 MILLIARIUM,
ROMAN MILE
1,616.66667 (a) yards
1,000.00000 (e) passus
41.66667 (r) acti
8.00000 (e) stadia
0.918561 (r) statute mile
1.4782829 (r) kilometers

The unit is also considered to have equaled 1,620 yards, 1,665 yards, 1,482 meters, and 500 pedes.

Linear units of Japan

Japan, as is true of most countries, is currently using the metric system, but the following units persist in many measuring instances.

1 MO
0.100000 (e) rin
0.001200 (a) inch
0.0304801 (r) millimeter

1 RIN	10.00000 (e) mo
	0.100000 (e) bu
	0.012000 (a) inch
	0.3048006 (r) millimeter
1 BU, BOO	100.00000 (e) mo
	10.00000 (e) rin
	0.120000 (a) inch
	0.100000 (e) sun
	0.3048006 (r) centimeter
1 SUN	1.19300 (a) inches
	0.100000 (e) shaku
	0.3048006 (r) decimeter
1 SHAKU,	10.00000 (e) sun
KUJIRA	0.994200 (a) feet
	0.100000 (e) jo
	0.3048006 (r) meter

The shaku is also considered to equal 1.243 feet.

1 HIRO	5.00000 (e) shaku
	4.97100 (a) feet
	1.5240030 (a) meters

The hiro is also considered to equal 1.515 meters.

1 KEN	6.00000 (e) shaku
	1.98800 (a) yards
	1.8288036 (r) meters
1 JO	10.00000 (e) shaku
	3.31400 (a) yards
	3.0480060 (r) meters
1 CHO	119.28000 (a) yards
	60.00000 (e) ken
	109.7282160 (r) meters

1 RI	36.00000 (e) cho
	2.43980 (a) statute miles
	39.5021578 (r) hectometers

A marine ri equals 1.15 statute miles.

Linear units of Russia

Russia, as is true of most countries, is currently using the metric system, but the following units persist in many measuring instances in Russia, Estonia, and Latvia.

| 1 TOTCHKA | 0.010000 (a) inch |
| | 0.2540005 (r) millimeter |

1 LINIYA	10.00000 (e) totchkas
	0.100000 (e) duim
	0.2540005 (r) centimeter

1 PALETZ	50.00000 (e) totchkas
	0.500000 (a) inch
	12.7000250 (r) millimeters

1 DUIM, DJUIM,   1.00000 (a) inch
DUIME, LIIN,   2.5400050 (r) centimeters
TOLL

The djuim is peculiar to the region of the Byelorussian Soviet Socialist Republic, as the liin and toll are to Estonia.

1 VERCHOK,   175.00000 (e) totchkas
VERCHOC   4.4450088 (r) centimeters

The unit is prevalent in Estonia and Latvia.

1 FOUTE, FUT   0.142857 (r) sagene
   30.4800600 (r) centimeters

---

1 ELLE   21.00000 (a) inches
                    53.3401050 (r) centimeters

                    The elle is peculiar to Estonia.

1 ARSHIN   27.96000 (a) inches
                    16.00000 (e) verchoks
                    0.333333 (r) sagene
                    71.1201400 (r) centimeters

                    The arshin is used in Russia, Estonia, and Latvia.

1 SAGENE,   7.00000 (e) foutes
SAGEN, FADEN,   6.99000 (a) feet
SULD   3.00000 (e) arshins
                    2.33000 (a) yards
                    21.3360420 (r) decimeters

                    The suld and faden are peculiar to Estonia, while the
                    sagene is used in Russia, Estonia, and Latvia.

1 VERST,   1,500.00000 (e) arshins
VERSTA, VERSTE 500.00000 (e) sagenes
                    0.661930 (a) statute mile
                    1.0652752 (r) kilometers

                    The unit is used in Russia, Estonia, and Latvia, and is
                    also considered to equal 0.663 statute mile; the verst
                    in Finland equals 0.664 statute mile.

Linear units of Spain

Spain, as is true of most countries, is currently
using the metric system, but the following units
persist in many measuring instances.

1 LINEA   0.076200 (a) inch
                    0.1935484 (r) centimeter

---

1 DEDO
9.00000 (e) lineas
0.685800 (a) inch
1.7419354 (r) centimeters

1 PULGADA
12.00000 (e) lineas
1.33333 (r) dedos
0.914400 (a) inch
0.083333 (r) pie
2.3225806 (r) centimeters

The pulgada of Mexico equals 0.916 inch.

1 SESMA
6.00000 (e) pulgadas
5.48640 (a) inches
13.9354836 (r) centimeters

1 CUARTA
9.00000 (e) pulgadas
8.22960 (a) inches
20.9032254 (r) centimeters

This cuarta is also used in Guatemala.

1 PIE
12.00000 (e) pulgadas
10.97280 (a) inches
27.8709696 (r) centimeters

The pie in both Mexico and Spain is also considered
to equal 10.992 inches and 27.86 centimeters.

1 CODO
16.45920 (a) inches
3.00000 (e) sesmas
0.500000 (e) vara
41.8064508 (r) centimeters

1 VARA
432.00000 (e) lineas
48.00000 (e) dedos
32.91840 (a) inches
6.00000 (e) sesmas

2.00000 (e) codos
0.500000 (e) estado
83.6129016 (r) centimeters

The Spanish vara is also considered to equal 32.88 inches and 32.913 inches, and has many other values as described under Linear Measure in Part One.

1 PASO
54.86510 (a) inches
1.66667 (r) varas
1.52403 (a) yards
139.3576231 (r) centimeters

1 ESTADO, BRAZA
65.83680 (a) inches
2.00000 (e) varas
1.82880 (a) yards
1.6722580 (r) meters

1 CORDEL
22.65045 (a) feet
8.25694 (r) varas
5.00000 (e) pasos
69.6788116 (r) decimeters

The cordel is an obsolete unit of old Spain and is also considered to have equaled 22.83 feet.

1 LEGUA
6,666.66667 (r) varas
3.46365 (a) statute miles
5.5742154 (r) kilometers

Linear units of Sweden

Sweden, as is true of most countries, is currently using the metric system, but the following units persist in many measuring instances.

1 LINJE
0.818050 (a) inch
2.0778511 (r) centimeters

The linje is also considered to equal 0.117 inch and 2.083 centimeters and 2.97 centimeters.

1 TUM

1.16800 (a) inches
0.100000 (e) fot
2.9667258 (r) centimeters

The tum is also considered to equal 1.12 inches.

1 FOT

11.68000 (a) inches
10.00000 (e) tums
0.500000 (e) aln
0.166667 (r) famn
0.100000 (e) stang
29.6672584 (r) centimeters

The unit is also considered to equal 11.2 inches and 29.69 centimeters.

1 ALN

23.36000 (a) inches
2.00000 (e) fotter
5.9334517 (r) decimeters

1 FAMN

6.00000 (e) fotter
5.84000 (a) feet
17.8003550 (r) decimeters

1 STANG

10.00000 (e) fotter
9.73333 (a) feet
2.9667258 (r) meters

The stang is also considered to equal 9.74 feet.

1 REF

100.00000 (e) fotter
97.33333 (a) feet
10.00000 (e) stang
29.6672584 (r) meters

The ref is also considered to equal 97.41 feet.

1 SWEDISH MILE   6.21371 (r) statute miles
10.0000000 (e) kilometers

In old Sweden, however, the unit equaled 6.64 statute miles.

Linear units of Switzerland

Switzerland, as is true of most countries, is currently using the metric system, but the following units persist in many measuring instances.

1 LINIE, LIGNE   0.082020 (a) inch
2.0833333 (r) millimeters

1 POUCE, ZOLL   1.18110 (r) inches
3.0000000 (e) centimeters

1 FUSS,   11.81103 (r) inches
SCHUN, PIED   10.00000 (e) pouces
0.250000 (e) staab
0.100000 (e) perche
30.0000000 (e) centimeters

1 STAAB, AUNE   47.24412 (r) inches
4.00000 (e) fuss
1.31234 (r) yards
12.0000000 (e) decimeters

The aune as defined above is also used in Belgium, while an aune in Jersey equals 48 inches and in old Paris equaled 46.79 inches.

1 KLAFTER   6.00000 (e) fuss
1.96851 (r) yards
18.0000000 (e) decimeters

1 PERCHE   10.00000 (e) fuss
9.84252 (r) feet

3.28084 (r) yards
3.0000000 (e) meters

1 LIEUE, STUNDE   1,600.00000 (e) perches
2.98259 (r) statute miles
4.8000000 (e) kilometers

The lieue of France equals 4.0 (e) kilometers.

Additional ancient and foreign linear units

Most of the countries or territories represented below now use the metric system, but the units mentioned still persist in many measuring instances.

ARGENTINA   **1 vara:** 2.842 (a) feet, 0.5 (e) braza, 8.6624331 (r) decimeters

1 braza: 5.684 (a) feet, 2.0 (e) varas, 17.3248661 (r) decimeters

1 klafter: 6.21 (a) feet, 18.9281173 (r) decimeters

1 cuadra: 426.3 (a) feet, 150.0 (e) varas, 75.0 (e) brazas, 68.64734 (a) klafters, 12.9936497 (r) decameters; also in Chile, 150 varas and 137.13 yards; in Ecuador, 100 varas and 91.42 yards; in Paraguay, 100 varas and 94.71 yards; and in Uruguay, 100 varas and 93.42 yards

BRAZIL   **1 pe:** 1.08 (a) feet, 3.2918465 (r) decimeters

1 covado: 26.96 (a) inches, 2.24667 (a) feet, 2.08333 (r) pes, 68.4785348 (r) centimeters

1 passo: 5.4 (a) feet, 5.0 (e) pes, 2.4 (r) covados, 16.4592324 (r) decimeters

1 braca: 7.18934 (a) feet, 3.2 (r) covados, 1.33333 (r) passos, 21.9131311 (r) decimeters; also considered to equal 7.22 feet and also used in Portugal

CHINA   **1 ts'un (new):** 1.41 (a) inches, 0.1 (e) ch'ih

1 ch'ih (new): 1.175 (a) feet, 0.01 (e) yin, 3.5814071

**China
(continued)**

(a) decimeters; also varies from 11.0 to 15.8 inches, but for customs purposes is 14.1 inches

1 pu (new): 70.5 (a) inches, 5.0 (e) ch'ih, 1.7907036 (a) meters

1 yin (old): 117.4976 (a) feet, 100.0 (e) ch'ih, 3.5814071 (a) or 3.5813339 (a) decameters

1 li (old): 2,114.9568 (a) feet, 18.0 (e) yin, 64.4640102 (a) decameters; used prior to China's conversion to metrics, equaled 1/1000 ch'ih, 1,800.0 ch'ih, 1/100 mu, and was also known as a Chinese mile

1 li (new) is equal to both 1.0 (e) millimeter and 1.0 (e) kilometer, and is also used as a surface measure equaling 1.0 (e) centaire

1 mu (old): 100.0 (e) li (old), 64.4640102 (a) kilometers; also occurs under Ancient and Foreign Surface Units in Part Two

1 tu (old): 4,500.0 (e) yin, 161.1600254 (a) kilometers; also equals 250.0 old li or 1/250 old li, as the occasion dictates

DENMARK

1 fod: 1.02917 (r) feet, 0.5 (e) alen, 3.1369163 decimeters

1 alen: 2.05833 (r) feet, 2.0 (e) fodder, 6.2738327 decimeters

1 favn: 6.0 (e) fodder, 6.17502 (r) feet, 3.0 (e) alen, 1.8821498 (r) meters

1 rode: 10.29167 (r) feet, 10.0 (e) fodder, 5.0 (e) alen, 3.1369163 (r) meters; also considered to equal 3.138 meters

EGYPT

1 theb: 0.738 (a) inch, 1.8745237 (r) centimeters; an ancient and obsolete unit

1 choryos: 75.0 (a) centimeters, 40.0 (e) thebs, 29.95276 (a) inches; an ancient and obsolete unit

1 abdat: 3.6 to 4.9 inches .

1 kassabah: 139.7664 (a) inches, 3.8824 (a) yards, 35.5007355 (r) decimeters

1 khet: 40.0 (e) (ancient) cubits, 23.2944 (a) yards, 6.0 (a) kassabah, 21.3004413 (a) meters; an ancient and obsolete unit, also considered to have equaled 23.0 yards

ICELAND

1 fet: 12.36 (a) inches, 0.5 (e) alin, 3.1394462 (r) decimeters

1 alin: 24.72 (a) inches, 2.0 (e) fet, 6.2788924 (r) decimeters

1 fathmur: 74.16 (a) inches, 6.0 (e) fet, 3.0 (e) alin, 1.883668 (r) meters

INDIA

1 jaob or jow: 0.25 (r) inch, 0.635001 (r) centimeter

1 guz: 144.0 (e) jaob, 1.0 (r) yard, 9.144018 (r) decimeters

1 danda, coss: 288.0 (e) jaob, 2.0 (e) guz, 1.8288036 (r) meters; primarily limited to Calcutta

1 yojan, yojana: 3,600.0 (a) dandas, 4.0909 (a) statute miles, 6.583693 (a) kilometers

ITALY

1 punto: 0.14 (a) inch, 0.3556007 (r) decimeter

1 braccio: 15 to 39 inches

1 canna: 39 to 118 inches

POLAND

1 cal: 1/12 stopa, 0.945 in, 24 millimeters

1 stopa: 11.34 (a) inches, 28.8036567 (r) centimeters

1 lokiec: 22.68 (a) inches, 2.0 (e) stopas, 5.7607313 (r) decimeters

1 sazen: 68.04 (a) inches, 6.0 (e) stopas, 3.0 (e) lokiec, 17.282194 (r) decimeters

1 pret: 15.0 (e) stopas, 4.725 (a) yards, 43.2054851 (r) decimeters

SIAM

1 anukabiet: 0.102526 (r) inch, 0.2604166 (r) centimeter

1 kabiet: 2.0 (e) anukabiet, 0.205052 (r) inch, 0.5208331 (r) centimeter; also considered to equal 0.206 inch

1 niu: 4.0 (e) kabiet, 0.820208 (r) inch, 2.0833333 (r) centimeters

1 keup: 12.0 (e) niu, 9.84253 (r) inches, 0.5 (e) (metric) sok, 25.0 (e) centimeters; the old keup equaled 10.0 inches and 1/16 wah

Siam
(continued)

1 (metric) sok: 19.68505 (r) inches, 2.0 (e) keup, 50.0 (e) centimeters; the old sok equaled 20.0 inches

1 ken: 39.3701 (r) inches, 2.0 (e) (metric) sok, 1.0 (e) meter; the old ken equaled 40.0 inches

1 wah, wa: 768.0 (e) anukabiet, 384.0 (e) kabiet, 78.74 (a) inches, 8.0 (e) keup, 4.0 (e) (metric) sok, 200.0 (e) centimeters; the old wah equaled 80.0 inches and 16 keup

1 sen: 366.63 (a) yards, 167.62 (a) wah, 335.24 (a) meters; an old and obsolete unit

1 roeneng: 2,000.0 (e) wah, 4.0 (e) kilometers

1 yut, yote: 400.0 (e) sen, 134.096 (a) kilometers, 83.325 (a) statute miles; also considered to equal 10.1 statute miles

YUGOSLAVIA

1 palaz: 1.42944 (r) inches, 3.6307848 (r) centimeters

1 rif: 30.59002 (r) inches, 21.4 (r) palaz, 7.7698794 (r) decimeters

1 khvat: 74.65 (r) inches, 1.8961137 (r) meters

MISCELLANEOUS FOREIGN UNITS

1 ady (*Malabar*): 10.46 inches

1 alen (*Norway*): 24.7 inches, 23.36 inches, 2.0 fotter

1 asta (*Malacca*): ⅛ jumba, 18.0 inches

1 bandle (*Ireland*): 2.0 feet

1 brasse (*France*): 1.62 meters

1 canna (*Malta*): 2.28 yards

1 chek (*Hong Kong*): 14⅝ inches

1 cordel (*old Cuba*): 66.77 feet

1 cordel (*old Paraguay*): 83⅓ varas, 76.42 yards

1 coss (*India*): 1.829 kilometers

1 covado (*Portugal*): 2.0 pes, 25.98 inches

1 dain (*Rangoon*): 2.43 statute miles

1 depa, depoh (*Dutch East Indies*): 8.0 kilan, 66.93 inches

1 douzieme (*watchmaker's unit*): 1/12 line, 0.007 inch

1 endere (*old Rumania*): 26.06 inches

1 estadio (*Portugal*): 282.0 yards

1 farsakh, farsang (*ancient Abyssinia*): 3.15 statute miles

1 **farsakh** (*Arabia*): ⅛ marhala, 3.0 statute miles

1 **farsakh, parasang** (*Persia*): 3.88 statute miles

1 **fotter** (*Norway*): 11.68 inches

1 **goad** (*obsolete, England*): 4.5 feet

1 **halibiu** (*Rumania*): 27.6 inches

1 **jumba** (*Malacca*): 8.0 astas, 144.0 inches

1 **kilan** (*Dutch East Indies*): ⅛ depa, 8.37 inches

1 **kup** (*Thailand*): 10.0 inches

1 **lan** (*Rangoon*): 2.24 yards

1 **latro** (*Czechoslovakia*): 2.096 yards

1 **loket** (*Czechoslovakia*): 0.579, 0.593, and 0.594 meter; a varying unit

1 **marhala** (*Arabia*): 8.0 farsakh, 24.0 statute miles

1 **marok** (*Hungary*): 4.15 inches

1 **me cate** (*Hondurus*): 24.0 varas, 20.0 meters

1 **mkono** (*East Africa*): 1.5 feet

1 **nin** (*Thailand*): 0.833 inches

1 **ona** (*Dominican Republic*): 1.3 yards

1 **punkt** (*Austria*): 1/728 fuss, 0.72 inches

1 **remen** (*ancient Egypt*): 14.6 inches, 37.0 centimeters

1 **rute, ruthe** (*Bavaria*): 10.0 fuss, 9.58 feet

1 **rute, ruthe** (*Prussia*): 12.0 fuss, 12.36 feet

1 **sah** (*Czechoslovakia*): 2.07 yards

1 **thou** (*United Kingdom*): 0.001 inches, 0.0254 millimeter

PART TWO

Surface units

Surface units are presented in Part Two:

Square measure

Comparative areas of square measure units

Units of ancient Rome

Units of Denmark, Iceland, Japan, Spain, Yugoslavia, and other foreign units, both current and ancient

Angular measure

For the purpose of these sections, *surface measures*, except angular measures, may be thought of as two-dimensional units consisting of length times width, or two units of length multiplied one by the other.

Square measure

1 SHED

10^{-21} (e) millibarn

10^{-24} (e) barn

10^{-48} (e) square centimeter

The shed is used in nuclear physics to measure cross sections of subatomic particles.

1 MILLIBARN

10^{+21} (e) sheds

0.001000 (e) barn

10^{-27} (e) square centimeter

The millibarn is used in nuclear physics to measure cross sections of subatomic particles. The term millibarn occurs also under Metric Surface Units in Part Five.

1 BARN, FERMI

10^{+24} (e) sheds

1,000.00000 (e) millibarns

10^{-24} (e) square centimeter

The barn is used in nuclear physics to measure cross sections of subatomic particles. The term barn also occurs under Metric Surface Units in Part Five; the term fermi also occurs under Linear Measure in Part One and under Metric Length Units and Metric Surface Units in Part Five.

1 CIRCULAR MIL

500,710,133,567,198,814,115.74789 (r) barns

0.776101 (r) square mil

0.000001 (e) circular inch

0.0005007 (r) square millimeter

The circular mil is defined as the area of a circle 1/1000 inch in diameter.

1 SQUARE MIL

6,451,625,400,025^{+8} (r) barns

1.28849 (r) circular mils

0.000001 (e) square inch

0.0006452 (r) square millimeter

1 SQUARE LINE
0.006944 (r) square inch
4.4802954 (r) square millimeters

1 CIRCULAR INCH
500,710,133,567,198,814,115,747,890.94211 (r) barns
1,000,000.00000 (e) circular mils
776,100.45223 (r) square mils
0.776101 (r) square inch
500.7112925 (r) square millimeters

1 SQUARE INCH
6,451,625,400,025^{+14} (r) barns
1,288,490.34371 (r) circular mils
1,000,000.00000 (e) square mils
144.00000 (e) square lines
1.28849 (r) circular inches
0.006944 (r) square foot
6.4516254 (r) square centimeters

The square inch is also considered to equal 6.45165 square centimeters.

1 SQUARE LINK
62.72640 (e) square inches
0.000100 (e) square chain
404.6872355 (r) square centimeters

1 SQUARE FOOT
20,736.00000 (e) square lines
144.00000 (e) square inches
0.111111 (r) square yard
0.003669 (r) perch
9.2903406 (r) square decimeters

1 SQUARE YARD
1,296.00000 (e) square inches
9.00000 (e) square feet
0.360000 (e) pace
0.090000 (e) square
0.033025 (r) perch
83.6130652 (r) square decimeters

1 SQUARE SCOTTISH ELL
1,369.00000 (e) square inches
9.50694 (r) square feet

0.027778 (r) fall

88.3227520 (r) square decimeters

1 FLAG   The flag is a common construction term indicating a 5.0-square-foot section of a concrete sidewalk.

1 PACE   25.00000 (e) square feet

2.77778 (r) square yards

0.250000 (e) square

2.3225852 (r) centares

The term pace also occurs under Linear Measure in Part One.

1 ROLL   30.00000 (e) square feet

3.33333 (r) square yards

2.7871022 (r) centares

The roll is used to measure wallpaper. The term also occurs under Paper Measure in Part Six.

1 SQUARE   36.00000 (e) square feet
FATHOM   4.00000 (e) square yards

3.3445226 (r) centares

The square fathom is used in the United States in the mining industry.

1 SQUARE   100.00000 (e) square feet

11.11111 (r) square yards

4.00000 (e) paces

9.2903406 (r) centares

The square is used to measure flooring and roofing.

1 PERCH,   625.00000 (e) square links
SQUARE PERCH,   272.25000 (e) square feet
SQUARE ROD,   30.35000 (e) square yards
SQUARE POLE   0.025000 (e) rood

0.007367 (r) arpent

1 *Perch,*	0.006250 (e) acre
square perch,	25.2929522 (r) centares
square rod,	
square pole	The unit is primarily used in surveying. The term perch also occurs under Linear Measure and Surveyor's Measure in Part One and under Cubic Measure in Part Three.
(continued)	

1 FALL

342.25000 (e) square feet
38.02778 (r) square yards
36.00000 (e) square Scottish ells
31.7961907 (r) centares

The fall is a Scottish unit.

1 SQUARE CHAIN

10,000.00000 (e) square links
4,356.00000 (e) square feet
484.00000 (e) square yards
16.00000 (e) perches
0.400000 (e) rood
0.117869 (r) arpent
0.100000 (e) acre
4.0468724 (r) ares

The square chain is the Gunter's or surveyor's chain squared, and is used in surveying. The unit is known as a ngan in Siam.

1 ROOD

10,890.00000 (e) square feet
1,210.00000 (e) square yards
40.00000 (e) perches
2.50000 (e) square chains
0.250000 (e) acre
10.1171809 (r) ares

The rood is a British unit used in surveying. An obsolete unit of Scotland, the particate, equaled 13,690 square feet and ¼ acre, and the maal or mal of Norway equals 10.0 (e) ares and ¼ acre.

1 ARPENT

36,956.21760 (e) square feet
4,106.24640 (e) square yards

135.74368 (r) perches

8.48398 (r) square chains

3.39359 (r) roods

0.848398 (r) acre

34.3335845 (r) ares

The arpent is a French unit used in surveying. Units similar to it include the morgen, juchert, and tagwerk or tagwert, all of Bavaria and equal to 0.842 (r) acre and 34.07 (r) ares.

The juchert or juchart of Switzerland, however, equals 0.89 (r) acre and 36 (e) ares, and the juchart and tagwerk of Württemberg equal 576 (e) square ruten, 1.168 (r) acres, and 47.27 (r) ares. The morgen of Prussia equaled 0.631 (r) acre and 25.53 (r) ares, while the morgen of South Africa, a morning's plowing, equals 2.116 (a) acres and 85.65 (a) ares.

1 ACRE

43,560.00000 (e) square feet

4,840.00000 (e) square yards

160.00000 (e) perches

10.00000 (e) square chains

4.00000 (e) roods

1.17869 (r) arpents

0.786990 (r) Scottish acre

0.617347 (r) Irish acre

0.472656 (r) Cheshire acre

0.000174 (r) section

40.4687236 (r) ares

Henry VIII of England limited the size of an acre to a piece of land 40 measuring rods long by 4 rods broad, each rod being about 5.5 yards long. The ancient Romans had used 10-foot rods, with each foot being about 13 inches.

1 SCOTTISH ACRE

6,150.00000 (e) square yards

161.72379 (r) falls

1.27066 (r) acres

51.4220351 (r) ares

The Scottish acre is also considered to equal 6,104 (e) square yards.

1 YOKE, JOCH   6,872.80000 (a) square yards

1.42000 (a) acres

57.4655875 (r) ares

The yoke is an Austrian unit. Units similar in name or size include the jutro or katastarsko of Yugoslavia, which equals 1.422 acres, and the jitro of Czechoslovakia, which equals 1.422 acres and 2.0 (e) koreci.

The hold, an agricultural unit of Hungary, equals 1.422 acres and 1.07 acres. The lanaz of Yugoslavia equals 1,600 (e) square khvats, 1.42 acres, and 57.5 ares. The sulung of old England equaled 4.0 (e) English yokes or jugums and 10.0 ares. The jugum of old England equaled 40.0 acres and 16.2 hectares.

1 IRISH ACRE   7,840.00000 (e) square yards

1.61984 (r) acres

65.5526431 (r) ares

1 CHESHIRE   10,240.00000 (e) square yards

ACRE   2.11570 (r) acres

85.6197788 (r) ares

The Cheshire acre is an old English unit, now obsolete.

1 SQUARE   10.00000 (e) acres

FURLONG   404.6872355 (r) ares

1 BLOCK,   The block or square is a varying unit denoting an

SQUARE   urban parcel of land bounded by streets. In Texas, however, a block equals 1.0 (e) square statute mile. The term square also occurs earlier in this section.

1 LOT   80.00000 (e) acres

3,237.4978839 (r) ares

1 HIDE   100.00000 (e) acres

4,046.8723549 (r) ares

1 QUARTER   160.00000 (e) acres

SECTION,   6,474.9957678 (r) ares

HOMESTEAD

1 LABOR	177.1400 (a) acres
	7,168.629592 (a) ares

The labor is used almost exclusively in Texas; in old Mexico a labor equaled 70.22 hectares.

1 SECTION,	2,560.00000 (e) roods
SQUARE	754.36160 (r) arpents
STATUTE MILE	640.00000 (e) acres
	0.111111 (r) square league
	0.027778 (r) township
	258.9998310 (r) hectares

1 BARONY	4,000.00000 (e) acres
	1,618.7489440 (r) hectares

The barony is an obsolete American and British unit.

1 SQUARE	5,760.00000 (e) acres
LEAGUE	9.00000 (e) sections
	0.250000 (e) township
	23.3099848 (r) square kilometers

1 TOWNSHIP	23,040.00000 (e) acres
	36.00000 (e) sections
	4.00000 (e) square leagues
	93.2399392 (r) square kilometers

Comparative areas of square measure units

The following table is presented to help the reader visualize the actual sizes of the various surface units. The figures given apply to the sides of squares and the radii of circles, each square and circle having an area equivalent to that of the unit it illustrates.

	Side		Radius	
	Inches	*Centimeters*	*Inches*	*Centimeters*
Square line	0.083(r)	0.212(r)	0.047(r)	0.120(r)
Square inch	1.000(e)	2.540(r)	0.564(r)	1.433(r)
Square link	7.920(e)	20.117(r)	4.468(r)	11.349(r)
Square foot	12.000(e)	30.480(r)	6.772(r)	17.201(r)
Square yard	36.000(e)	91.440(r)	20.316(r)	51.603(r)
Pace, flag	60.000(e)	152.400(r)	33.864(r)	86.015(r)

	Side		Radius	
	Feet	*Decimeters*	*Feet*	*Decimeters*
Roll	5.478(r)	16.697(r)	3.091(r)	9.421(r)
Square fathom	6.000(e)	18.288(r)	3.386(r)	10.321(r)
Square	10.000(e)	30.480(r)	5.643(r)	17.191(r)
Perch	16.500(e)	50.292(r)	9.311(r)	28.380(r)
Square chain	66.000(e)	201.168(r)	37.242(r)	113.514(r)
Rood	˙104.367(r)	318.111(r)	58.893(r)	179.506(r)
Arpent	192.240(r)	585.998(r)	108.479(r)	330.644(r)
Acre	208.710(r)	636.148(r)	117.773(r)	358.672(r)
Scottish acre	235.266(r)	717.091(r)	132.759(r)	404.647(r)
Yoke	248.706(r)	758.056(r)	140.316(r)	427.683(r)
Irish acre	265.632(r)	809.640(r)	149.892(r)	456.871(r)
Cheshire acre	303.582(r)	925.318(r)	171.309(r)	522.150(r)

	Side		Radius	
	Yards	*Meters*	*Yards*	*Meters*
Lot	196.774(r)	179.931(r)	111.040(r)	101.535(r)
Hide	220.000(e)	201.168(r)	124.146(r)	113.519(r)
Quarter section	880.000(e)	804.672(r)	496.584(r)	454.275(r)
Section	1,760.000(e)	1,609.344(r)	993.149(r)	908.136(r)
Barony	4,400.000(e)	4,023.360(r)	2,482.920(r)	2,270.382(r)
Square league	5,280.000(e)	4,828.032(r)	2,979.680(r)	2,724.619(r)

Surface units of ancient Rome

The following units may prove useful to the archeologist and to students of the arts and history.

1 SQUARE PES
0.940901 (a) square foot
0.104545 (a) square yard
0.001000 (e) scrupulus
8.7412816 (r) square decimeters

The square pes is also considered to have equaled 0.01046 square yard.

1 SCRUPULUS,
SCRUPULUM
1,000.00000 (e) square pedes
104.54445 (a) square yards
0.277778 (r) clima
0.021600 (a) acre
87.4128157 (r) square meters

The unit is also considered to have equaled 10.46 square yards and 8.74 square meters.

1 CLIMA
3,600.00000 (e) square pedes
3.600000 (e) scrupuli
0.750000 (e) actus simplex
0.077760 (a) acre
0.062500 (e) heredium
31.4686136 (r) ares

The clima is also considered to have equaled 3.15 ares.

1 ACTUS
SIMPLEX
4,800.00000 (e) square pedes
501.81314 (a) square yards
4.80000 (e) scrupuli
1.33333 (r) climae
0.480000 (e) versus
0.342856 (r) actus major
0.200000 (e) uncia
0.103680 (a) acre
419.5815154 (r) square meters

1 Actus simplex (*continued*)	The actus simplex is also considered to have equaled 50.2 square yards.

1 VERSUS

10,000.00000 (e) square pedes
1,054.44445 (a) square yards
10.00000 (e) scrupuli
2.77778 (r) climae
2.08333 (r) acti simplices
0.714286 (r) actus major
0.416667 (r) uncia
0.216001 (a) acre
8.7412443 (r) ares

1 ACTUS MAJOR

14,000.00000 (e) square pedes
1,463.62223 (a) square yards
14.00000 (e) scrupuli
3.88889 (r) climae
2.91667 (r) acti simplices
1.40000 (e) versus
0.583333 (r) uncia
0.486111 (r) juger
0.302401 (a) acre
12.2377942 (r) ares

The actus major is also considered to have equaled 0.31 acre and 12.6 ares.

1 UNCIA

24,000.00000 (e) square pedes
2,530.66668 (a) square yards
5.00000 (e) acti simplices
2.40000 (e) versus
1.71429 (r) acti majores
0.833333 (r) juger
0.518402 (a) acre
0.416667 (r) heredium
20.9790758 (r) ares

The uncia is also considered to have equaled 251 square yards and 210 square meters.

1 JUGER, **JUGERUM**	28,800.00000 (e) square pedes 2.88000 (e) versus 2.05714 (r) acti majores 1.20000 (e) unciae 0.622083 (a) acres 0.500000 (e) heredium 25.1748909 (r) ares
1 HEREDIUM	57,600.00000 (e) square pedes 16.00000 (e) climae 4.11428 (r) acti majores 2.40000 (e) unciae 2.00000 (e) jugera 1.24417 (a) acres 0.010000 (e) centuria 50.3497812 (r) ares
1 CENTURIA	5,760,000.00000 (e) square pedes 200.00000 (e) jugera 124.41653 (a) acres 100.00000 (e) heredia 0.250000 (e) saltus 50.3497812 (r) hectares
1 SALTUS	23,040,000.00000 (e) square pedes 800.00000 (e) jugera 497.66612 (a) acres 400.00000 (e) heredia 4.00000 (e) centuriae 201.3991274 (r) hectares

Surface units of Denmark

Denmark, as is true of most countries, is currently using the metric system, but the following units persist in many measuring instances.

---

1 SQUARE ALEN   0.3812072 (r) centare

1 ALBUM   748.96063 (r) square alens
                    11.66667 (r) square rods
                    0.333333 (r) fjerding
                    295.5093533 (r) centares

                    The album is also considered to equal 11.68 square
                    rods.

1 FJERDING   2,246.88188 (r) square alens
                    35.00000 (r) square rods
                    3.00000 (e) albums
                    0.218750 (r) acre
                    885.2806000 (r) centares

1 TONDELAND   14,000.00000 (e) square alens
                    1.31877 (a) acres
                    53.3690122 (r) ares

                    The tondeland is also considered to equal 1.363 acres.

1 TONDE   71,900.22000 (a) square alens
HARTKORN   1,120.00000 (r) square rods
                    96.00000 (e) albums
                    32.00000 (e) fjerdings
                    7.00000 (r) acres
                    5.13573 (r) tondelands
                    283.2897920 (r) ares

Surface units of Iceland

Iceland, as is true of most countries, is currently
using the metric system, but the following units
persist in many measuring instances.

1 FERTHUMI-   1.06000 (a) square inches
UNGUR   0.006944 (r) ferfet
                   6.8387229 (r) square centimeters

---

1 FERFET
144.00000 (e) ferthumiungurs
1.06000 (a) square feet
0.250000 (e) feralin
0.027778 (r) ferfathmur
9.8477610 (r) square decimeters

1 FERALIN
576.00000 (e) ferthumiungurs
4.24000 (a) square feet
4.00000 (e) ferfets
39.3910440 (r) square decimeters

1 FERFATHMUR
5,184.00000 (e) ferthumiungurs
36.00000 (e) ferfets
9.00000 (e) feralins
3.5451940 (r) square meters

The ferfathmur is also considered to equal 3.546 square meters.

1 TUNDAG-SLATTA
900.00000 (e) ferfathmurs
31.9067456 (r) ares

The tundagslatta is also considered to equal 31.914 ares.

1 ENGJATEIGUR
1,600.00000 (e) ferfathmurs
1.77778 (r) tundagslattas
56.7231034 (r) ares

The engjateigur is also considered to equal 56.74 ares.

1 FERMILA
10,000.00000 (e) engjateigurs
21.90079 (a) square statute miles
56.7231034 (a) square kilometers

Surface units of Japan

Japan, as is true of most countries, is currently using the metric system, but the following units persist in many measuring instances.

| 1 SHAKU | 0.988000 (a) square foot |
| | 9.1788565 (r) square decimeters |

1 GO	3.55680 (a) square feet
	0.100000 (e) bu
	33.0438835 (r) square decimeters

1 BU, TSUBO	36.00000 (e) shaku
	35.56800 (a) square feet
	10.00000 (e) go
	3.95200 (a) square yards
	3.3043884 (r) centares

1 SE	1,080.00000 (e) shaku
	1,067.04000 (a) square feet
	118.56000 (a) square yards
	30.00000 (e) bu
	0.100000 (e) tan
	99.1316503 (r) centares

1 TAN	10,800.00000 (e) shaku
	300.00000 (e) bu
	10.00000 (e) se
	0.244959 (r) acre
	0.100000 (e) cho
	9.9131650 (r) ares

1 CHO	3,000.00000 (e) bu
	10.00000 (e) tan
	2.449587 (r) acres
	99.1316503 (r) ares

The cho is also considered to equal 2.451 acres.

Surface units of Spain

Spain, as is true of most countries, is currently using the metric system, but the following units persist in many measuring instances.

1 SQUARE VARA   0.062500 (e) estadal

0.5941754 (r) centare

The term vara occurs also under Linear Measure and Linear Units of Spain in Part One.

1 ESTADAL   16.00000 (e) square varas

11.37000 (a) square yards

9.5068055 (r) centares

The estadal of Nicaragua equals 16 square varas and 11.34 square yards.

1 CELEMIN   768.00000 (e) square varas

545.76000 (a) square yards

0.1127603 (a) acre

48.00000 (e) estadals

456.3266646 (r) centares

The celemin is also considered to equal 0.133 acre.

1 ARANZADA   6,400.00000 (e) square varas

400.00000 (e) estadals

8.33333 (r) celemins

0.9396694 (a) acre

38.0272221 (r) ares

The aranzada is also considered to equal 1.105 acres.

1 FANEGADA,   9,216.00000 (e) square varas
MARCO REAL   6,549.12000 (a) square yards

12.00000 (e) celemins

1.35312 (a) acres

54.7591998 (r) ares

The unit is also considered to equal 1.591 acres. A fanegada in Venezuela equals 1.73 acres; in the Canary Islands, 1.3 acres; and in Peru, 64.5 ares.

1 YUGADA   72.00000 (e) aranzadas

27.3795999 (r) hectares

The yugada is also considered to equal 32.198 hectares.

1 CABALLERIA	81.1872 (a) acres
	60.00000 (e) fanegadas
	3.2855520 (r) hectares

The caballeria is also considered to equal 95.48 acres, and the unit is used in other countries as follows: In Costa Rica, the caballeria equals 64.5 manzanas and 111.82 acres; in Cuba, 33.162 acres; in El Salvador, 111.11 acres; in Guatemala, 64.5 manzanas and 111.51 acres; in Honduras, 64.5 manzanas and 111.13 acres; in Mexico, 12 fanegas and 105.75 acres; in Nicaragua, 64.5 manzanas and 112.41 acres; and in Puerto Rico, 194.1 acres.

Surface units of Yugoslavia

Yugoslavia, as is true of most countries, is currently using the metric system, but the following units persist in many measuring instances.

1 DONUM	837.18950 (r) square yards
	7.0000000 (e) ares

The donum in Cyprus equals 1,600 square yards; in Libya, 1,600 square piks and 1.83 acres; in Palestine, 9.0 ares; in old Turkey, 40 square piks and 0.277 acre; and in new Turkey, 25 ares.

1 MOT YKA	956.78800 (r) square yards
	1.14286 (r) donums
	8.0000000 (e) ares

1 RALICA, RALO	0.617775 (r) acre
	25.0000000 (e) ares

1 DAN ORANJA	1.44000 (e) ralicas
	0.889596 (r) acre
	36.0000000 (e) ares

Additional ancient and foreign surface units

Most of the countries or territories represented below now use the metric system, but the units mentioned still persist in many measuring instances.

EGYPT

1 aurure: 0.677 acre; an ancient unit now obsolete

1 feddan, feddan masri: 333⅓ kassabah, 1.038 acres; in ancient Arabia, 400 qasaba or achir and 1.47 acres

1 kassabah: 135.646 square feet

1 sahme: 1/576 feddan, 8.72 square yards

ITALY

1 giornata: 38.0 acres

1 quadrato: 1.25 acres

1 tavola: 1/100 giornata, 45 square yards

PARAGUAY

1 cuadra: 1.85 acres; in Chile, 22,500 square varas and 3.88 acres; in Peru, 2.47 acres; in Uruguay, 1.82 acres; in Brazil, 0.92 acre; and in Argentina, 22,500 square varas and 4.17 acres

1 legua: 1,875 hectares; the term legua also occurs under Linear Measure in Part One

1 line, lino: 75.0 ares; old and obsolete units; the lino also equaled 100 square varas. The term line also occurs under Linear Measure, Cloth Measure, and Printer's Measure in Part One, and under Electrical Units of Magnetic Flux in Part Eight.

MISCELLANEOUS FOREIGN UNITS

1 acaena (*ancient Greece*): 11.0 square yards

1 balita (*Philippines*): 1/10 quinon, 0.69 acre

1 bigha (*Bengal*): 1,600 square yards

1 bigha (*India, United Province*): 3,025 square yards

1 bouw (*Dutch East Indies*): 1.754 acres

1 bunder (*Netherlands*): 100.0 ares

1 cawney, cawny: (*Madras*): 1.322 acres

1 cordel (*old Cuba*): 1/324 caballeria, 0.102 acre

1 cotta, cottah (*Bengal*): 80 square yards

1 cover (*Wales*): 0.667 acre

1 cuerda, cuerdo (*Puerto Rico*): 0.971 acre

Miscellaneous foreign units (continued)

1 darat (*Somaliland*): 80 ares

1 deciatine (*Latvia*): 2.7 acres

1 dekare (*Bulgaria*): 0.247 acre

1 dessiatine (*Russia*): 2,400 square sagenes, 2.7 acres

1 fanega (*Mexico*): 8.81 acres

1 ferrado (*Portugal*): 605 square varas, 0.179 acre, 7.25 ares

1 gong qing (*Mainland China*): 2.471 acres

1 heminee (*old southern France*): 8 to 16 ares

1 jabia (*Libya*): 1,800 square piks, 995.4 square yards

1 jerib (*ancient Persia*): 1,000 to 1,066 square guz, 10.82 ares

1 joch (*Austria*): 1.422 acres

1 joch (*Hungary*): 1.067 acres

1 journée (*France*): 1.0 acre (a)

1 kappland (*Sweden*): 1,750 square fotter, 184.5 square yards

1 kish (*China*): ¼ mu, 183.7 square yards

1 lan (*Czechoslovakia*): 60 koreci, 42.67 acres

1 loan (*Philippines*): 1/10 balita, 2.79 ares

1 lofstelle (*Estonia*): 0.458 acre

1 lofstelle (*Latvia*): 0.918 acre

1 manzana (*Costa Rica, El Salvador*): 1.727 acres

1 manzana (*Honduras*): 1.723 acres

1 manzana (*Nicaragua*): 1.742 acres

1 merice (*Bohemia*): 20.0 ares

1 mira, korec, staych (*Czechoslovakia*): 28.78 ares

1 mishara (*Iraq*): 0.618 acre

1 morg, morga (*Poland*): 300 square prets, 55.99 ares

1 mu (*Mainland China*): 0.165 acre

1 mud (*Netherlands*): 2.471 acre

1 plethron, plethrum (*ancient Greece*): 10,000 square podes, 0.235 acre

1 pouruete (*Latvia*): 0.92 acre

1 quarteron (*France*): 1,076 square meters

1 quinon (*Philippines*): 10 balitas, 2.795 hectares

1 quo (*Annam*): 1,800 square ngu, 1.07 hectares

1 rai (*Siam*): 16 ares

1 ropani (*Nepal*): 94.05 square meters
1 sao (*Annam*): 9 square ngu, 64 square yards
1 se (*Japan*): 118.615 square yards
1 sitio (*Mexico*): 492.28 fanegas, 1755.61 hectares
1 suerte (*Nicaragua*): 1.41 hectares
1 suerte (*Uruguay*): 2,700 cuadras, 7.69 square statute miles
1 tarea (*Cuba*): 82.56 square yards
1 tarea (*Dominican Republic*): 0.155 acre
1 tonne (*Schleswig-Holstein, West Germany*): 1.35 acres
1 tonnstelle (*Latvia*): 1.286 acres
1 topo (*Peru*): 0.669 acre
1 tun (*Estonia*): 1.093 hectares
1 tunland, tunnland (*Finland*): 1.22 acres
1 vloka, wloka (*Poland*): 16.796 hectares

Angular measure

Angular or circular measure, like other surface measures, may also be thought of as planimetry, the study of plane areas, first devised by the Babylonians, based on a circular earth, and now used extensively for nautical and air travel needs, as well as a multitude of other uses.

1 SECOND
0.016667 (r) minute
0.000494 (r) mil
0.000278 (r) degree

The term second also occurs under Time Measure in Part Six.

1 MINUTE,
MINUTE OF
THE ARC
60.00000 (e) seconds
0.029630 (r) mil
0.016667 (r) degree
0.001481 (r) point

The term minute also occurs under Time Measure in Part Six.

1 MIL
202.50000 (e) seconds
3.37500 (e) minutes
0.562500 (e) degree
0.005000 (e) point
0.001875 (r) sign
0.000982 (r) radian

The mil is used primarily in gunnery measure. The term also occurs under Linear Measure in Part One.

1 CENTRAD
0.010000 (e) radian
0.001592 (r) circle

1 DEGREE
3,600.00000 (e) seconds
60.00000 (e) minutes
1.77778 (r) mils
0.088889 (r) point
0.033333 (r) sign
0.017453 (r) radian
0.016667 (r) sextant
0.011111 (r) quadrant
0.002778 (r) circle

The term degree also occurs under Linear Measure in Part One, Temperature Units in Part Seven, and Electrical Cycle Units in Part Eight.

1 QUARTER, QUARTER POINT
168.75000 (e) minutes
0.250000 (e) point

The unit is described as one fourth the distance between any two adjacent points of the 32 marked on a compass, that is, 2 degrees, 48 minutes, 45 seconds.

1 POINT
40,500.00000 (e) seconds
675.00000 (e) minutes
200.00000 (e) mils
11.25000 (e) degrees
4.00000 (e) quarters
0.375000 (e) sign
0.196349 (r) radian

0.187500 (e) sextant

0.125000 (e) quadrant

0.031250 (e) circle

The point is a marine unit used in navigation. The term also occurs under Linear Measure and Printer's Measure in Part One and under Troy Weight in Part Four.

1 HOUR

The hour is a unit of right ascension, the degree of the equator which, in the right sphere, rises at the same moment with any celestial body. Now usually it is the distance eastward or counterclockwise along the celestial equator, from the first point of Aries to the hour circle passing through any celestial body. Analogously to terrestrial longitude, right ascension is expressed sometimes in degrees, but usually in time, and is most conveniently measured by noting the time elapsing between the culmination of the vernal equinox and of the body in question. The hour in this sense represents 15 degrees, or one twenty-fourth of a great circle. The term hour also occurs under Time Measure in Part Six.

1 SIGN

1,800.00000 (e) minutes

533.33333 (r) mils

30.00000 (e) degrees

2.66667 (r) points

0.523600 (r) radian

0.500000 (e) sextant

0.333333 (r) quadrant

0.083333 (r) circle

1 OCTANT

162,000.00000 (e) seconds

0.125000 (e) circle

1 RADIAN

206,264.80000 (r) seconds

3,437.75000 (r) minutes

1,018.59200 (r) mils

57.29578 (r) degrees

5.09296 (r) points

1.90986 (r) signs

1 Radian	0.954931 (r) sextant
(continued)	0.636620 (r) quadrant
	0.159155 (r) circle

The radian is defined as that angle of a circle sub-tended by an arc equal in length to the circle's radius; hence, 1 radian equals 180 degrees divided by pi, or 57 degrees, 17 minutes, 44.8 seconds.

A steradian or spheradian is that solid angle of a sphere subtended by a portion of the surface whose area is equal to the square of the sphere's radius.

1 SEXTANT

3,600.00000 (e) minutes
1,066.66667 (r) mils
60.00000 (e) degrees
5.33333 (r) points
2.00000 (e) signs
1.04720 (r) radians
0.666667 (r) quadrant
0.166667 (r) circle

1 QUINTANT

72.00000 (e) degrees
0.20000 (e) circle

1 SEMICIRCLE, HEMICYCLE

0.500000 (e) circle

1 CIRCLE, REVOLUTION, CIRCUMFERENCE

1,296,000.00000 (e) seconds
21,600.00000 (e) minutes
6,400.00000 (e) mils
628.31855 (r) centrads
360.00000 (e) degrees
32.00000 (e) points
12.00000 (e) signs
8.00000 (e) octants
6.28318 (r) radians
4.00000 (e) quadrants

The term revolution also occurs under Electrical Cycle Units in Part Eight.

PART THREE

Units of capacity and volume

Capacity and volume units are presented in Part Three:

Cubic measure

Liquid measure

Dry measure

Apothecaries' fluid measure

Cooking measure

Spirits measure

Comparative capacities and volumes of cubic, liquid, dry, apothecaries' fluid, cooking, and spirits measures

Intermeasure comparison of capacity and volume units

Units of ancient Arabia, Greece, India, Israel, and Rome

Units of Argentina, Austria, Brazil, Cyprus, Denmark, Egypt, France, Iceland, Japan, Libya, Mexico, the Netherlands, Poland, Portugal, Russia, Spain, Sweden, Switzerland, and other foreign units, both ancient and current

Volume is defined as the space occupied by anything, but the *contents* can be influenced by temperature and gravity. For example, when the white of an egg is heated, its volume increases, as does the volume of water that is heated and becomes steam.

Volume and *capacity* imply a three-dimensional enclosure, length by width by height, of any shape; most units described in Part Three are thought of in terms of capacity, a size of unvarying and uninfluenced dimensions.

The first exact measure of capacity was used in ancient Babylon, its name lost in antiquity; it was a hollow cube, each side measuring a square handbreadth, and the amount of water it held was accepted as a unit of capacity.

Cubic measure

1 MIL-FOOT,
CIRCULAR
MIL-FOOT

0.000009 (r) cubic inch
154.4480782 (r) milliliters
0.1544482 (r) cubic millimeter

The mil-foot is an electrical unit used to measure wire size; 1 mil-foot is equivalent to a wire measuring 1 mil in diameter by 1 foot in length.

1 CUBIC INCH,
INCH CUBE

106,078.32171 (r) mil-feet
0.000579 (r) cubic foot
16.3870640 (r) deciliters
0.0163872 (r) cubic decimeter

The cubic inch is also considered to equal 0.016372 cubic decimeter.

1 BOARD FOOT

144.00000 (e) cubic inches
0.083333 (r) cubic foot
2,359.7372160 (e) deciliters
2.3597510 (r) cubic decimeters

The board foot is used in measuring lumber. Its legal dimensions are 1 foot square by 1 inch thick.

1 CUBIC FOOT   1,728.00000 (e) cubic inches
TIMBER FOOT   12.00000 (e) board feet
　　　　　　　　　0.200000 (e) bulk barrel
　　　　　　　　　0.062500 (e) cord foot
　　　　　　　　　0.040404 (r) perch
　　　　　　　　　0.037037 (r) cubic yard
　　　　　　　　　0.028571 (r) displacement ton
　　　　　　　　　0.025000 (e) marine ton
　　　　　　　　　28.3162196 (r) liters
　　　　　　　　　28.3170125 (r) cubic decimeters

The cubic foot is called a load if used to measure earth; the term load is used again in this section and also occurs under Dry Measure in Part Three.

1 BULK BARREL   5.00000 (e) cubic feet
　　　　　　　　　0.185185 (r) cubic yard
　　　　　　　　　0.142857 (r) displacement ton
　　　　　　　　　0.125000 (e) marine ton
　　　　　　　　　0.050000 (e) register ton
　　　　　　　　　0.006250 (e) load
　　　　　　　　　141.5810980 (r) liters
　　　　　　　　　141.5850624 (r) cubic decimeters

The bulk barrel is used in shipping. A barrel of petroleum is considered to equal 42.0 (e) U.S. gallons or 158.86 liters.

1 CORD FOOT   192.00000 (e) board feet
　　　　　　　　　16.00000 (e) cubic feet
　　　　　　　　　0.646465 (r) perch
　　　　　　　　　0.592593 (r) cubic yard
　　　　　　　　　0.148148 (r) stack
　　　　　　　　　453.0595140 (r) liters
　　　　　　　　　453.0722000 (r) cubic decimeters

1 STANDARD   200.00000 (e) board feet
　　　　　　　　　16.66667 (r) cubic feet
　　　　　　　　　0.101010 (r) British standard

1 Standard	471.9369934 (r) liters
(continued)	471.9502080 (r) cubic decimeters

1 PERCH

297.00000 (e) board feet
24.75000 (e) cubic feet
1.54688 (r) cord feet
0.916667 (r) cubic yard
0.229167 (r) stack
700.8540239 (r) liters
700.8736478 (r) cubic decimeters

The legal dimensions of a perch, as used in the measure of brick or stone, are 16.5 feet long by 1.5 feet wide by 1 foot high. The term perch also occurs under Linear and Surveyor's Measure in Part One and under Square Measure in Part Two.

1 CUBIC YARD

324.00000 (e) board feet
27.00000 (e) cubic feet
5.40000 (e) bulk barrels
1.68750 (e) cord feet
1.09091 (r) perches
0.771429 (r) displacement ton
0.675000 (e) marine ton
0.250000 (e) stack
0.210938 (r) cord
0.077143 (r) keel
0.033750 (e) load
764.5680261 (r) liters
764.5894340 (r) cubic decimeters

The cubic yard is also considered to equal 0.7645594 cubic meter.

1 DISPLACEMENT TON

35.00000 (e) cubic feet
7.00000 (e) bulk barrels
1.29630 (r) cubic yards
0.875000 (e) marine ton
0.350000 (e) register ton

0.100000 (e) keel
0.043750 (e) load
0.005000 (e) shipload
991.1067000 (r) liters
991.1344510 (r) cubic decimeters

The displacement ton is used in marine shipping to indicate the amount of water a ship displaces, a displacement ton of sea water weighing 1.0 long ton.

1 MARINE TON, CARGO TON, MEASUREMENT TON, FREIGHT TON, SHIPPING TON, TIMBER TON

480.00000 (e) board feet
40.00000 (e) cubic feet
8.00000 (e) bulk barrels
2.50000 (e) cord feet
1.61616 (r) perches
1.48148 (r) cubic yards
1.14286 (r) displacement tons
0.400000 (e) register ton
0.370370 (r) stack
0.312513 (r) cord
0.114286 (r) keel
0.050000 (e) load
1,132.6933720 (r) liters
1,132.7250874 (r) cubic decimeters

The marine ton was once considered to equal 11/25 cubic yard or 38.88 (e) cubic feet.

1 BRITISH SHIPPING TON

42.00000 (e) cubic feet
1,189.2812232 (r) liters
1,189.3145250 (r) cubic decimeters

1 TIMBER LOAD

600.00000 (e) board feet
50.00000 (e) cubic feet
1.85185 (r) cubic yards
0.303030 (r) British standard
1,415.8667150 (r) liters
1,415.9063600 (r) cubic decimeters

The timber load is a British unit.

1 STANDARD HUNDRED

The standard hundred is basically a unit of count, of pieces of cut lumber, encompassing various sizes of boards; standard industry amounts of these boards equal various cubic capacities, which in turn are measured in deals or thills, as boards so used are called. The standard hundred is divided into four different categories to supply the building industry with the most used board sizes:

1 St. Petersburg equals 120 deals or thills, each board measuring 1.5 inches by 11.0 inches by 12.0 feet, the total amount of wood equaling 1,980 board feet

1 Christiana or **Christiania** equals 120 deals, each board measuring 1.25 inches by 9.0 inches by 11.0 feet, the total equaling 1,237.5 board feet

1 Quebec equals 100 deals, each board measuring 3.0 inches by 11.0 inches by 10.0 feet, the total equaling 27.5 cubic feet

1 London equals 120 deals, each board measuring 3.0 inches by 9.0 inches by 12.0 feet, the total equaling 17.0 cubic feet

The deal or thill, today always of fir or pine, is itself cut to any of several specified sizes. For instance, a standard deal, from which the others are sawed, is usually 3 inches by 9 inches by 12 feet, as in the London category. A whole deal indicates a thickness of 1.25 inches; a split deal indicates 5⁄8 inch thickness, and five-cut stuff is 0.5 inch or less thick. Pieces less than 6.0 feet long are called deal ends.

In the United States export trade, yellow pine is sawed 9.0 inches and wider, and 3, 4, or 5 inches thick. A Quebec deal is a piece of northern pine at least 3.0 inches thick and of any width; a Quebec standard deal measures 3.0 inches by 11.0 inches by 12.0 feet; and a St. Petersburg standard deal measures 1.5 inches by 11.0 inches by 12.0 feet.

Other "standard lengths" of the lumber industry are: in rough pine lumber, multiples of 2 feet, from 4 feet to 24 feet inclusive; in hardwoods, from 4 feet to 16 feet; and in surface lumber, from 3 feet to 20 feet, including the odd lengths.

1 REGISTER TON

100.00000 (e) cubic feet

20.00000 (e) bulk barrels

3.70370 (r) cubic yards

2.85714 (r) displacement tons
2.50000 (e) marine tons
0.285714 (r) keel
0.125000 (e) load
0.014286 (r) shipload
2,831.7334300 (r) liters
2,831.8127200 (r) cubic decimeters

The register ton is used in marine shipping to indicate how much cargo a ship will hold. It is known as a tonneau de mer in Belgium and a tonneau de jauge in international measure.

1 STACK

108.00000 (e) cubic feet
6.75000 (e) cord feet
4.36364 (r) perches
4.00000 (e) cubic yards
2.70000 (e) marine tons
0.843750 (e) cord
0.135000 (e) load
3,058.2721044 (r) liters
3,058.3577360 (r) cubic decimeters

The stack is a British unit used in the measure of wood.

1 CORD

128.00000 (e) cubic feet
8.00000 (e) cord feet
5.17167 (r) perches
4.74074 (r) cubic yards
3.20000 (e) marine tons
1.18519 (r) stacks
0.365714 (r) keel
0.160000 (e) load
3,624.6187904 (r) liters
3,624.7202797 (r) cubic decimeters

The cord is used in the measure of wood. Its legal dimensions are 8.0 feet in length by 4.0 feet in width by 4.0 feet in height.

1 BRITISH   165.00000 (e) cubic feet
STANDARD   3.30000 (e) timber loads
9.90000 (e) standards
4,672.2796877 (r) liters
4,672.4105115 (r) cubic decimeters

1 KEEL   350.00000 (e) cubic feet
70.00000 (e) bulk barrels
12.96293 (r) cubic yards
10.00000 (e) displacement tons
8.75000 (e) marine tons
3.50000 (e) register tons
0.437500 (e) load
0.050000 (e) shipload
9,911.0670050 (r) liters
9,911.3445149 (r) cubic decimeters

The term keel also occurs under Avoirdupois Weight in Part Four.

1 LOAD   800.00000 (e) cubic feet
160.00000 (e) bulk barrels
50.00000 (e) cord feet
32.32323 (r) perches
29.62963 (r) cubic yards
22.85714 (r) displacement tons
20.00000 (e) marine tons
8.00000 (e) register tons
7.40741 (r) stacks
6.25000 (e) cords
2.29143 (r) keels
0.114286 (r) shipload
22,653.8674400 (r) liters
22,654.5017483 (r) cubic decimeters

A load also equals 1 cubic foot if used to measure earth. The term load is also used in Dry Measure, Part Three, and in Avoirdupois Weight, Part Four.

1 ACRE-INCH
3,630.00000 (e) cubic feet
102.7878772 (r) decaliters
102,790.7553750 (r) cubic decimeters

The acre-inch is defined as the volume of water required to cover 1 acre of land 1 inch deep. The measure is used in relation to irrigation.

1 SHIPLOAD
7,000.00000 (e) cubic feet
350.00000 (e) bulk barrels
259.25926 (r) cubic yards
200.00000 (e) displacement tons
175.00000 (e) marine tons
70.0000 (e) register tons
20.00000 (e) keels
8.75000 (e) loads
198.2213401 (r) decaliters
198,226.8902975 (r) cubic decimeters

The shipload is used in the measure of coal and coke.

1 ACRE-FOOT
43,560.00000 (e) cubic feet
12.00000 (e) acre-inches
1,233.5030821 (r) kiloliters
1,233.5376202 (r) cubic meters

The acre-foot is defined as the volume of water required to cover 1 acre of land 1 foot deep. The measure is used in relation to irrigation.

**1 CUBIC
STATUTE MILE**
32,768,000.00000 (e) cubic rods
3,379,200.00000 (e) acre-feet
4.1680677 (r) cubic kilometers

Liquid measure

Henry VII of England in the fifteenth century declared that "eight pounds do make a gallon of wine, and eight gallons of wine do make a bushel," a decree that contributed to the creation of dry and liquid measuring systems.

1 GILL,
QUARTERN

0.832647 (r) imperial gill
0.250000 (e) pint
0.1182908 (r) liter
0.1182941 (r) cubic decimeter

The term quartern also occurs under Dry Measure in Part Three and under Avoirdupois Weight in Part Four. The term gill originally meant "one drink of wine."

1 IMPERIAL
GILL, NOGGIN

1.20099 (r) gills
0.333333 (r) mutchkin
0.250000 (e) imperial pint
0.1420636 (r) liter
0.1420686 (r) cubic decimeter

The unit is British. The term noggin is also used in Spirits Measure, Part Three.

1 MUTCHKIN

3.00000 (e) imperial gills
0.750000 (e) imperial pint
0.4261937 (r) liter
0.4262057 (r) cubic decimeter

1 PINT

4.00000 (e) gills
0.832596 (r) imperial pint
0.500000 (e) quart
0.125000 (e) gallon
0.4731632 (r) liter
0.4731765 (r) cubic decimeter

The pint, the word probably originating from medieval Latin meaning "painted mark" on a measuring container, is also considered to equal 0.473167 liter. The term pint also occurs under Dry Measure, Apothecaries' Fluid Measure, Cooking Measure, and Spirits Measure in Part Three.

1 IMPERIAL PINT   4.00000 (e) imperial gills
1.33333 (r) mutchkins
1.20099 (r) pints
0.500000 (e) imperial quart
0.125000 (e) imperial gallon
0.5682583 (r) liter
0.5682742 (r) cubic decimeter

1 QUART   8.00000 (e) gills
2.00000 (e) pints
0.832596 (r) imperial quart
0.500000 (e) pottle
0.250000 (e) gallon
0.9463264 (r) liter
0.9463529 (r) cubic decimeter

The word quart originally meant a "quarter part," and the unit is known as a chupak in Malaysia. The term quart is also used in Dry Measure, Cooking Measure, and Spirits Measure in Part Three.

1 IMPERIAL QUART   8.00000 (e) imperial gills
2.00000 (e) imperial pints
1.20099 (r) quarts
0.250000 (e) imperial gallon
1.1365166 (r) liters
1.1365485 (r) cubic decimeters

1 POTTLE, STOUP   4.00000 (e) pints
3.33043 (r) imperial pints
2.00000 (e) quarts
0.500000 (e) gallon
1.8926528 (r) liters
1.8927058 (r) cubic decimeters

The pottle was originally an old English unit. The term pottle also occurs under Dry Measure in Part Three.

1 GALLON	32.00000 (e) gills
	8.00000 (e) pints
	4.00000 (e) quarts
	2.00000 (e) pottles
	0.832596 (r) imperial gallon
	0.031746 (r) barrel
	3.7853056 (r) liters
	3.7854116 (r) cubic decimeters

The gallon is known as a gantang in Sabah. The term gallon also occurs under Apothecaries' Fluid Measure, Cooking Measure, and Spirits Measure in Part Three, and under Avoirdupois Weight in Part Four.

1 IMPERIAL GALLON	9.60760 (r) pints
	8.00000 (e) imperial pints
	4.80380 (r) quarts
	4.00000 (e) imperial quarts
	1.20099 (r) gallons
	0.222222 (r) pin
	0.038127 (r) barrel
	4.5460665 (r) liters
	4.5461938 (r) cubic decimeters

The imperial gallon is defined as the volume of 10.0 pounds of pure water, at 62 degrees Fahrenheit, and containing 277.418 cubic inches.

1 PIN	36.00000 (e) imperial pints
	21.61775 (r) quarts
	18.00000 (e) imperial quarts
	10.80887 (r) pottles
	5.40446 (r) gallons
	4.50000 (e) imperial gallons
	0.171564 (r) barrel
	0.120000 (e) cran
	0.011250 (e) tank
	20.4573003 (r) liters
	20.4578721 (r) cubic decimeters

The pin is also considered to equal ⅛ barrel.

1 BARREL

126.00000 (e) quarts
31.50000 (e) gallons
26.22844 (r) imperial gallons
0.500000 (e) hogshead
119.2371264 (r) liters
119.2404650 (r) cubic decimeters

The barrel is also considered to equal 43.2 gallons, and 1 barrel of petroleum is considered to equal 42.0 gallons or 158.86 liters. The term barrel also occurs under Dry Measure and Spirits Measure in Part Three and under Avoirdupois Weight in Part Four.

1 CRAN

150.00000 (e) imperial quarts
45.03713 (r) gallons
37.50000 (e) imperial gallons
8.33333 (r) pins
0.093750 (e) tank
170.4774945 (r) liters
170.4822575 (r) cubic decimeters

The cran is a British unit used in the measure of fresh herring; originally, the Gauls used the word *crann* to mean "a lot."

1 DRUM

The drum varies from 50 to 55 gallons in capacity.

1 HOGSHEAD

63.00000 (e) gallons
52.45689 (r) imperial gallons
2.00000 (e) barrels
238.4742528 (r) liters
238.4809301 (r) cubic decimeters

The term hogshead also occurs under Spirits Measure in Part Three and under Avoirdupois Weight in Part Four. A similar unit, the fanega of Paraguay, equals 63.4 gallons.

1 WATER TON

The water ton is an old and obscure British unit equaling 1,018.3262432 (a) liters and 62,127.367 (a) cubic inches.

1 TANK	480.39600 (a) gallons
	400.00000 (e) imperial gallons
	88.88889 (r) pins
	10.66667 (r) crans
	1,818.4266080 (a) liters
	1,818.4775200 (a) cubic decimeters

Dry measure

1 PINT	0.500000 (e) quart
	0.250000 (e) pottle
	0.062500 (e) peck
	0.031250 (e) bucket
	0.015625 (e) bushel
	0.015140 (r) British bushel
	0.5506105 (r) liter
	0.5506259 (r) cubic decimeter

The pint is also considered to equal 0.550595 (r) liter and 0.550599 (r) liter. The term pint also occurs under Liquid Measure, Apothecaries' Fluid Measure, Cooking Measure, Spirits Measure in Part Three.

1 BASKET   Farmers use long climax baskets, which equal from 0.5 (e) pint to several quarts. Apples generally come in round stave baskets of bushel size. Splint baskets, usually holding strawberries, equal a pint or quart.

1 QUART	2.00000 (e) pints
	0.500000 (e) pottle
	0.125000 (e) peck
	0.062500 (e) bucket
	0.031250 (e) bushel
	0.030280 (r) British bushel
	0.009524 (r) barrel
	1.1012210 (r) liters
	1.1012518 (r) cubic decimeters

The quart is also considered to equal 1.101198 liters. The term quart also occurs under Liquid Measure, Cooking Measure, and Spirits Measure in Part Three.

1 POTTLE,
QUARTERN,
LIPPY,
STIMPART

4.00000 (e) pints
2.00000 (e) quarts
0.250000 (e) peck
0.125000 (e) bucket
0.062500 (e) bushel
0.060559 (r) British bushel
2.2024419 (r) liters
2.2025036 (r) cubic decimeters

The term pottle also occurs under Liquid Measure in Part Three; the term quartern also occurs under Liquid Measure in Part Three and under Avoirdupois Weight in Part Four.

1 PECK

16.00000 (e) pints
8.00000 (e) quarts
4.00000 (e) pottles
0.500000 (e) bucket
0.250000 (e) bushel
0.242234 (r) British bushel
0.125000 (e) strike
0.083333 (r) bag
8.8097676 (r) liters
8.8100143 (r) cubic decimeters

The peck is also considered to equal 8.80958 liters. The term also occurs under Cooking Measure in Part Three.

1 BRITISH PECK

1.03147 (a) pecks
9.0870000 (a) liters
9.0875298 (a) cubic decimeters

1 BUCKET

32.00000 (e) pints
16.00000 (e) quarts
8.00000 (e) pottles
2.00000 (e) pecks
0.500000 (e) bushel
0.484469 (r) British bushel
0.250000 (e) strike

1 Bucket	0.166667 (r) bag
(continued)	0.142857 (r) windle
	0.062500 (e) quarter
	17.6195351 (r) liters
	17.6200285 (r) cubic decimeters

1 BUSHEL	64.00000 (e) pints
	32.00000 (e) quarts
	16.00000 (e) pottles
	4.00000 (e) pecks
	2.00000 (e) buckets
	0.968937 (r) British bushel
	0.500000 (e) strike
	0.333333 (r) bag
	0.304762 (r) barrel
	0.285714 (r) windle
	0.125000 (e) quarter
	35.2390702 (r) liters
	35.2400569 (r) cubic decimeters

The bushel is also considered to equal 0.35238 hectoliters. The word *bushel*, from old Celtic usage, meant "the hollowed hand." An ancient Egyptian unit, the heket, the word roughly meaning "a sack" in the measure of gold, is similar now to the bushel.

The term bushel also occurs under Cooking Measure in Part Three and under Avoirdupois Weight in Part Four.

1 FIRLOT	1.03096 (r) bushels
	0.250000 (e) boll
	36.3300000 (a) liters
	36.3310532 (a) cubic decimeters

The firlot is a Scottish unit.

1 BRITISH	66.04995 (r) pints
BUSHEL,	33.02498 (r) quarts
ENGLISH	16.51249 (r) pottles
IMPERIAL	4.12812 (r) pecks
BUSHEL	2.06406 (r) buckets

1.03203 (r) bushels
0.516015 (r) strike
0.344010 (r) bag
0.294866 (r) windle
0.250000 (e) coombe
0.129004 (r) quarter
0.025000 (e) load
36.3687944 (r) liters
36.3698128 (r) cubic decimeters

The British bushel equals 8.0 imperial gallons and 2,219.36 cubic inches, as defined above; however, it is also considered to equal 2,218.192 cubic inches.

An obsolete British unit, the Winchester bushel or struck bushel, equaled 2,178.0 cubic inches, and a heaped bushel equaled 1.25 Winchester bushels.

1 STRIKE

128.00000 (e) pints
64.00000 (e) quarts
32.00000 (e) pottles
8.00000 (e) pecks
4.00000 (e) buckets
2.00000 (e) bushels
1.93787 (r) British bushels
0.666667 (r) bag
0.571429 (r) windle
0.484469 (r) coombe
0.250000 (e) quarter
70.4781404 (r) liters
70.4801138 (r) cubic decimeters

The strike is also considered to vary from 2.0 pecks to 4.0 bushels.

1 BAG

192.00000 (e) pints
96.00000 (e) quarts
48.00000 (e) pottles
12.00000 (e) pecks
6.00000 (e) buckets
3.00000 (e) bushels

1 Bag
(continued)

2.90681 (r) British bushels
1.50000 (e) strikes
0.857144 (r) windle
0.726704 (r) coombe
105.7201707 (r) liters
105.7231309 (r) cubic decimeters

The bag is also known as a sack, a British unit also considered to equal 109 (a) liters. The terms bag and sack also occur under Avoirdupois Weight in Part Four.

1 BARREL

210.00000 (e) pints
105.00000 (e) quarts
52.50000 (e) pottles
13.12500 (e) pecks
6.56250 (e) buckets
3.28125 (e) bushels
3.17941 (r) British bushels
1.64063 (r) strikes
1.09375 (e) bags
0.936607 (r) windle
0.794853 (r) coombe
0.410156 (r) quarter
115.6282050 (r) liters
115.6314426 (r) cubic decimeters

The barrel equals 7,056 (a) cubic inches when used as above, but equals 5,826 (a) cubic inches when used to measure cranberries. The term barrel also occurs under Liquid Measure and Spirits Measure in Part Three and under Avoirdupois Weight in Part Four.

1 WINDLE

224.00000 (e) pints
112.00000 (e) quarts
56.00000 (e) pottles
14.00000 (e) pecks
7.00000 (e) buckets
3.50000 (e) bushels
3.39128 (r) British bushels

1.75000 (e) strikes
1.16667 (r) bags
1.06667 (r) barrels
0.847820 (r) coombe
0.437500 (e) quarter
123.3401995 (r) liters
123.3436530 (r) cubic decimeters

The windle is a British unit used in measuring wheat.

1 BOLL

4.12384 (r) bushels
4.00000 (e) firlots
145.3200000 (a) liters
145.3240690 (a) cubic decimeters

The boll in Scotland and North England equals 1/16 chalder and 2 to 6 bushels; ⅛ boll in Scotland is called an auchlet and varies from ¼ to ¾ bushels and from 9 to 27 liters. One chalder in Scotland equals 64.0 (e) Edinburgh firlots, 65.9 bushels, and 23.22 hectoliters.

**1 COOMBE,
COOMB, COOM,
COMB**

264.19980 (r) pints
132.09990 (r) quarts
66.04995 (r) pottles
16.51249 (r) pecks
8.25624 (r) buckets
4.12812 (r) bushels
4.00000 (e) British bushels
2.06410 (r) strikes
1.37604 (r) bags
1.17946 (r) windles
0.516015 (r) quarter
0.100000 (e) load
0.050000 (e) last
145.4751776 (r) liters
145.4792509 (r) cubic decimeters

The unit is British and is also considered to equal 1.0 boll and 2.0 strikes.

1 ARDEB

The ardeb is primarily an Egyptian unit officially equivalent to 5.62 bushels, but varies considerably in other neighboring countries.

1 QUARTER,
SEAM

512.00000 (e) pints
256.00000 (e) quarts
128.00000 (e) pottles
32.00000 (e) pecks
16.00000 (e) buckets
8.00000 (e) bushels
7.75148 (r) British bushels
4.00000 (e) strikes
2.66667 (r) bags
2.28571 (r) windles
1.93787 (r) coombes
0.193787 (r) load
0.096894 (r) last
281.9204552 (r) liters
281.9283490 (r) cubic decimeters

The term seam also occurs under Avoirdupois Weight in Part Four. The term quarter also occurs under Cloth Measure in Part One and under Troy Weight and Avoirdupois Weight in Part Four. The quarter is also considered to equal ¼ chauldron and is used in the measure of coal.

1 CHAULDRON

The chauldron equals 4.0 quarters of coal; varies from 32 to 36 British bushels in the measure of coal, lime, and coke; and is usually 32 bushels when used to measure wheat. The term chaldron occurs under Avoirdupois Weight in Part Four.

1 LOAD, WEY

660.49950 (r) pottles
165.12488 (r) pecks
82.56244 (r) buckets
41.28122 (r) bushels
40.00000 (e) British bushels
20.64061 (r) strikes
13.76041 (r) bags
11.79464 (r) windles

10.00000 (e) coombes
5.16015 (r) quarters
0.500000 (e) last
1,454.7925091 (r) liters
1,454.8332432 (r) cubic decimeters

The load is a British unit, and the term also occurs under Cubic Measure in Part Three and under Avoirdupois Weight in Part Four. The term wey also occurs under Avoirdupois Weight in Part Four.

1 LAST

330.24975 (r) pecks
165.12488 (r) buckets
82.56244 (r) bushels
80.00000 (e) British bushels
41.28122 (r) strikes
27.52081 (r) bags
23.58927 (r) windles
20.00000 (e) coombes
10.32031 (r) quarters
2.00000 (e) loads
2,909.5850182 (r) liters
2,909.6664864 (r) cubic decimeters

The last is a British unit used to measure grain. Its value varies with locality: in Belgium and Holland the last equals 85.134 bushels; in Prussia, 112.29 bushels. The term last also occurs under Avoirdupois Weight in Part Four.

1 CARLOAD

The carload is used in the United States in the measure of coal and equals 9 4/9 or 9.44444 chauldrons. In the measure of cereal, however, the carload equals 340 U.S. bushels and 119.8 hectoliters.

Apothecaries' fluid measure

**1 BRITISH
MINIM**

0.960782 (r) minim
0.911743 (r) drop
0.050000 (e) fluid scruple

1 British minim (*continued*)	0.016030 (r) fluid dram
	0.0591936 (r) milliliter

1 MINIM	1.04082 (r) British minims
	0.950000 (e) drop
	0.016667 (r) fluid dram
	0.0616098 (r) milliliter

1 DROP	1.09560 (r) British minims
	1.05263 (r) minims
	0.017544 (r) fluid dram
	0.0648524 (r) milliliter

The drop is also considered to equal 1 minim, but may vary with the specific gravity and viscosity of the liquid. The term drop also occurs under Cooking Measure in Part Three.

1 FLUID SCRUPLE	20.00000 (e) British minims
	19.21564 (r) minims
	18.23486 (r) drops
	0.320261 (r) fluid dram
	1.1838717 (r) milliliters

1 FLUID DRAM, FLUIDRAM, FLUID DRACHM, TEASPOON	60.00000 (e) minims
	57.00000 (e) drops
	3.12246 (r) fluid scruples
	0.800000 (e) English teaspoon
	0.500000 (e) dessert spoon
	0.250000 (e) tablespoon
	0.125000 (e) fluid ounce
	3.6965875 (r) milliliters

The term teaspoon is also used in Cooking Measure, Part Three.

1 ENGLISH TEASPOON	78.06150 (r) British minims
	75.00000 (e) minims
	1.25000 (e) fluid drams
	0.625000 (e) dessert spoon

0.312500 (e) tablespoon
0.250000 (e) English tablespoon
4.6207343 (r) milliliters

1 DESSERT
SPOON

120.00000 (e) minims
114.00000 (e) drops
2.00000 (e) fluid drams
1.60000 (e) English teaspoons
0.500000 (e) tablespoon
0.400000 (e) English tablespoon
0.250000 (e) fluid ounce
7.3931749 (r) milliliters

1 TABLESPOON

4.00000 (e) fluid drams
3.20000 (e) English teaspoons
2.00000 (e) dessert spoons
0.800000 (e) English tablespoon
0.500000 (e) fluid ounce
14.7863472 (r) milliliters

The term tablespoon is also used in Cooking Measure, Part Three.

1 ENGLISH
TABLESPOON

5.00000 (e) fluid drams
4.00000 (e) English teaspoons
2.50000 (e) dessert spoons
1.25000 (e) tablespoons
0.625000 (e) fluid ounce
18.4829373 (r) milliliters

1 BRITISH FLUID
OUNCE

480.00000 (e) British minims
461.17536 (r) minims
24.00000 (e) fluid scruples
7.68626 (r) fluid drams
6.14901 (r) English teaspoons
1.53725 (r) English tablespoons
0.960782 (r) fluid ounce
0.266884 (r) English breakfast cup

1 British *fluid ounce* *(continued)*	0.006250 (e) imperial gallon 0.0284129 (r) liter 0.0283871 (r) cubic decimeter

1 FLUID OUNCE
480.00000 (e) minims
456.00000 (e) drops
24.97968 (r) fluid scruples
8.00000 (e) fluid drams
6.40000 (e) English teaspoons
4.00000 (e) dessert spoons
2.00000 (e) tablespoons
1.60000 (e) English tablespoons
1.04082 (r) British fluid ounces
0.625000 (e) fluid pint
0.520400 (r) British pint
0.0295727 (r) liter
0.0295735 (r) cubic decimeter

The fluid ounce defined above equals the ounce used in Spirits Measure, Part Three.

1 WINE GLASS
960.00000 (e) minims
912.00000 (e) drops
16.00000 (e) fluid drams
8.00000 (e) dessert spoons
2.00000 (e) fluid ounces
0.666667 (r) teacup
0.0591454 (r) liter
0.0591470 (r) cubic decimeter

1 TEACUP
1,440.00000 (e) minims
1,368.00000 (e) drops
24.00000 (e) fluid drams
3.00000 (e) fluid ounces
1.50000 (e) wine glasses
0.833333 (r) English breakfast cup
0.0887181 (r) liter
0.0887205 (r) cubic decimeter

1 ENGLISH   1,798.53696 (r) British minims
BREAKFAST CUP   1,728.00000 (e) minims
                    3.74695 (r) British fluid ounces
                    3.60000 (e) fluid ounces
                    1.20000 (e) teacups
                    0.1064617 (r) liter
                    0.1064646 (r) cubic decimeter

1 FLUID PINT   7,680.00000 (e) minims
                    7,296.00000 (e) drops
                    128.00000 (e) fluid drams
                    16.00000 (e) fluid ounces
                    8.00000 (e) wine glasses
                    0.832640 (r) British pint
                    0.125000 (e) gallon
                    0.4731632 (r) liter
                    0.4731765 (r) cubic decimeter

The fluid pint equals the pint used in Liquid Measure, and Spirits Measure, Part Three.

1 BRITISH PINT 153.72672 (r) fluid drams
                    19.21584 (r) fluid ounces
                    9.60792 (r) wine glasses
                    1.20099 (r) fluid pints
                    0.150124 (r) gallon
                    0.5682583 (r) liter
                    0.5682742 (r) cubic decimeter

1 GALLON   1,024.00000 (e) fluid drams
                    128.00000 (e) fluid ounces
                    64.00000 (e) wine glasses
                    8.00000 (e) fluid pints
                    6.66112 (r) British pints
                    0.832640 (r) imperial gallon
                    3.7853056 (r) liters
                    3.7854116 (r) cubic decimeters

1 Gallon *(continued)*	The term gallon also occurs under Liquid Measure, Cooking Measure, and Spirits Measure in Part Three, and under Avoirdupois Weight in Part Four.
1 IMPERIAL **GALLON**	160.00000 (e) British fluid ounces 9.60792 (r) fluid pints 8.00000 (e) British pints 1.20099 (r) gallons 4.5460665 (r) liters 4.5461938 (r) cubic decimeters

The term imperial gallon is also used in Liquid Measure, Part Three.

Cooking measure

The following kitchen units are perhaps more precise than the average cook needs; the values are carried out to five, six, or seven places, to be shortened as the cook desires.

1 DROP	0.526316 (r) coffee spoon 0.0648524 (r) milliliter

The term drop is also used in Apothecaries' Fluid Measure, Part Three.

1 COFFEE **SPOON**	19.00000 (e) drops 0.250000 (e) teaspoon 0.083333 (r) tablespoon 1.2321956 (r) milliliters

1 TEASPOON	76.00000 (e) drops 4.00000 (e) coffee spoons 0.333333 (r) tablespoon 4.9287824 (r) milliliters

The teaspoon is also considered to equal 120.0 (e) drops and 60.0 (e) drops. The term teaspoons is also used in Apothecaries' Fluid Measure, Part Three.

1 BRITISH **TEASPOON**	95.00000 (e) drops 1.25000 (e) teaspoons 6.1609780 (r) milliliters

1 TABLESPOON	228.00000 (e) drops 12.00000 (e) coffee spoons 3.00000 (e) teaspoons 0.062500 (e) cup 0.031250 (e) pint (liquid) 14.7863472 (r) milliliters

Usually about 1 tablespoon of salt or 4 tablespoons of flour equals approximately 1 ounce. The term tablespoon is also used in Apothecaries' Fluid Measure, Part Three.

1 CUP	192.00000 (e) coffee spoons 48.00000 (e) teaspoons 16.00000 (e) tablespoons 0.500000 (e) pint (liquid) 0.250000 (e) quart (liquid) 0.062500 (e) gallon 0.2365816 (r) liter 0.2365882 (r) cubic decimeter

Usually about 2.0 cups of granulated sugar, 2.5 cups of confectioner's sugar, 2⅔ cups of brown sugar, 4.0 cups of flour, and 2.0 cups of chopped meat or butter equals approximately 1 avoirdupois pound.

1 (LIQUID) PINT	384.00000 (e) coffee spoons 96.00000 (e) teaspoons 32.00000 (e) tablespoons 2.00000 (e) cups 0.500000 (e) quart (liquid) 0.125000 (e) gallon 0.053709 (r) peck 0.013427 (r) bushel 0.4731632 (r) liter 0.4731765 (r) cubic decimeter

---

1 (Liquid) pint
(continued)

The term pint also occurs under Liquid Measure, Dry Measure, Apothecaries' Fluid Measure, and Spirits Measure in Part Three.

1 (DRY) PINT

0.500000 (e) quart (dry)
0.5506105 (r) liter
0.5506259 (r) cubic decimeter

The term pint also occurs under Liquid Measure, Dry Measure, Apothecaries' Fluid Measure, and Spirits Measure in Part Three.

1 (LIQUID) QUART

768.00000 (e) coffee spoons
192.00000 (e) teaspoons
64.00000 (e) tablespoons
4.00000 (e) cups
2.00000 (e) pints (liquid)
0.250000 (e) gallon
0.107418 (r) peck
0.026855 (r) bushel
0.9463264 (r) liter
0.9463529 (r) cubic decimeter

The term quart also occurs under Liquid Measure, Dry Measure, and Spirits Measure in Part Three.

1 (DRY) QUART

2.00000 (e) pints (dry)
0.125000 (e) peck
1.1012210 (r) liters
1.1012518 (r) cubic decimeters

The term quart also occurs under Liquid Measure, Dry Measure, and Spirits Measure in Part Three.

1 BRITISH CUP

10.00000 (e) cups
2.3658160 (r) liters
2.3658800 (r) cubic decimeters

The unit equals 8.0 fluid ounces of Apothecaries' Fluid Measure, Part Three.

---

1 GALLON

256.00000 (e) tablespoons
16.00000 (e) cups
8.00000 (e) pints (liquid)
4.00000 (e) quarts (liquid)
0.429671 (r) peck
0.107418 (r) bushel
3.7853056 (r) liters
3.7854116 (r) cubic decimeters

The term gallon also occurs under Liquid Measure, Apothecaries' Fluid Measure, and Spirits Measure in Part Three, and under Avoirdupois Weight in Part Four.

1 PECK

37.23776 (r) cups
18.61888 (r) pints (liquid)
16.00000 (e) pints (dry)
9.30944 (r) quarts (liquid)
8.00000 (e) quarts (dry)
2.32736 (r) gallons
0.250000 (e) bushel
8.8097676 (r) liters
8.8100143 (r) cubic decimeters

The term peck is also used in Dry Measure, Part Three.

1 BUSHEL

148.95102 (r) cups
74.47551 (r) pints (liquid)
64.00000 (e) pints (dry)
32.00000 (e) quarts (dry)
9.30944 (r) gallons
4.00000 (e) pecks
35.2390702 (r) liters
35.2400569 (r) cubic decimeters

The term bushel also occurs under Dry Measure in Part Three and under Avoirdupois Weight in Part Four.

Spirits measure

Spirits measure, useful to the bartender, cook, refreshment retailer, party host, and others, is the measure of beverages, primarily those with alcoholic content.

Possibly the smallest liquor bottles still currently marketed are the 24-minim bottles of Scotch whisky, which contain 1/20 fluid ounce, sold by the Cumbrae Supply Company of Scotland.

1 PONY

0.750000 (e) shot
0.500000 (e) jigger
0.0221795 (r) liter
0.0221801 (r) cubic decimeter

The pony also equals 1.0 fluid ounce, the same as the shot or finger, and additionally is considered to equal about 4.0 fluid ounces, especially when used to measure brandy.

1 SHOT, OUNCE

1.33333 (r) ponies
0.666667 (r) jigger
0.0295727 (r) liter
0.0295735 (r) cubic decimeter

The term ounce also occurs under Avoirdupois Weight and Apothecaries' Weight in Part Four. The spirits unit finger generally attempts to imitate a shot or ounce of whisky, being equal in container depth to the width of a man's finger.

1 JIGGER

2.00000 (e) ponies
1.50000 (e) shots
0.0443591 (r) liter
0.0443603 (r) cubic decimeter

1 DOUBLE

2.00000 (e) shots
0.0591454 (r) liter
0.0591470 (r) cubic decimeter

1 TRIPLE

3.00000 (e) shots
0.0887181 (r) liter
0.0887205 (r) cubic decimeter

1 NOGGIN,
IMPERIAL GILL

4.80388 (r) shots
0.1420636 (r) liter
0.1420686 (r) cubic decimeter

The noggin equals one "drink" of whisky. The terms noggin and imperial gill are also used in Liquid Measure, Part Three.

1 SPLIT

The split is a varying unit equal to half the commercial volume of a spirit, generally champagne or wine, and is construed as being a drink of one half the usual quantity.

1 (LIQUID)
PINT

21.33333 (r) ponies
16.00000 (e) shots
10.66667 (r) jiggers
0.625000 (e) fifth
0.500000 (e) quart (liquid)
0.250000 (e) magnum
0.4731632 (r) liter
0.4731765 (r) cubic decimeter

A tankard, mug, or stein is generally a container for ale or beer, and usually equals about a pint. The term pint also occurs under Liquid Measure, Dry Measure, Apothecaries' Fluid Measure, and Cooking Measure in Part Three.

1 QUARTER
YARD

1.25000 (a) pints (liquid)
0.5914540 (a) liter
0.5914700 (a) cubic decimeter

1 FIFTH

34.13333 (r) ponies
25.60000 (e) shots
17.06667 (r) jiggers
1.60000 (e) pints (liquid)
0.800000 (e) quart (liquid)

1 Fifth	0.400000 (e) magnum
(continued)	0.200000 (e) gallon
	0.7570611 (r) liter
	0.7570823 (r) cubic decimeter

1 BOTTLE (WINE)	0.800633 (r) quart (liquid)
	0.400317 (r) magnum
	0.250354 (r) jeroboam
	0.166667 (r) imperial gallon
	0.016667 (r) anker
	0.7576778 (r) liter
	0.7576990 (r) cubic decimeter

The wine bottle is also considered to equal 0.5 magnum and 0.8 liter. The boutylka, a wine unit of Russia, equals 1.625 liquid pints and 0.769 liter.

1 (LIQUID)	42.66667 (r) ponies
QUART	32.00000 (e) shots
	21.33333 (r) jiggers
	2.00000 (e) pints (liquid)
	1.25000 (e) fifths
	0.500000 (e) magnum
	0.250000 (e) gallon
	0.9463264 (r) liter
	0.9463529 (r) cubic decimeter

The term quart also occurs under Liquid Measure, Dry Measure, and Cooking Measure in Part Three.

1 HALF YARD	2.50000 (a) pints (liquid)
	2.00000 (e) quarter yards
	0.500000 (e) yard
	1.1829080 (a) liters
	1.1829400 (a) cubic decimeters

1 MAGNUM	4.00000 (e) pints (liquid)
	2.49797 (r) bottles (wine)
	2.00000 (e) quarts (liquid)

0.625000 (e) jeroboam
0.500000 (e) gallon
1.8926528 (r) liters
1.8927058 (r) cubic decimeters

The magnum besides equaling ⅖ gallon (and other values) also equals 1.514 liters.

1 YARD

4.00000 (e) quarter yards
2.00000 (e) half yards
0.512081 (r) ale gallon
2.3658160 (a) liters
2.3658800 (a) cubic decimeters

The yard is an old English unit used in the measure of ale and beer, which were served in a vessel 1 yard in height. The term yard also occurs under Linear Measure, Mariner's Measure, Surveyor's Measure, and Cloth Measure in Part One.

1 JEROBOAM

6.40000 (e) pints (liquid)
3.99670 (r) bottles (wine)
3.20000 (e) quarts (liquid)
1.60000 (e) magnums
0.800000 (e) gallon
3.0282445 (r) liters
3.0283293 (r) cubic decimeters

The jeroboam, named for Jeroboam I, king of ancient North Israel, is used mostly for brandy and champagne.

Monopole champagne (proprietary brands of wine shippers) is also marketed in bottle sizes, including the quarter bottle of 0.25 liter, half bottle of 0.5 liter, and the following:

The rehoboam, which equals 6 bottles or 3 magnums, 1 gallon 9 ounces or about 4.8 liters

The methuselah, which equals 8 bottles or 4 magnums, 1 gallon 1 quart 1 pint 5 ounces or about 6.4 liters

The salmanezer, which equals 12 bottles or 6 magnums, 2 gallons 18 ounces or about 9.6 liters

The balthazar, which equals 16 bottles or 8 magnums, 2 gallons 3 quarts 10 ounces or about 12.8 liters

1 Jeroboam *(continued)*	**The nebuchadnezzar,** which equals 20 bottles or 10 magnums, 4 gallons 1 pint 13 ounces or about 16 liters

For Bordeaux wine are included the imperial, equaling 8 bottles and about 6 liters, and the marie-jeanne of approximately 3 bottles or about 2.5 liters; for port wine, the tappit-hen, equaling 3 bottles or about 2.27 liters.

Perhaps the largest bottle ever made was the adelaide, a 5-foot-tall sherry bottle made in Stoke on Trent, Staffordshire, England, in 1958, with a capacity of 93.19 liters.

1 GALLON,
DOUBLE
MAGNUM

8.00000 (e) pints (liquid)

5.00000 (e) fifths

4.00000 (e) quarts (liquid)

2.00000 (e) magnums

1.25000 (e) jeroboams

0.832596 (r) imperial gallon

0.819330 (r) ale gallon

0.333038 (r) barn gallon

3.7853056 (r) liters

3.7854116 (r) cubic decimeters

A liquid gallon in Haiti equals 0.8326 of the gallon defined above, and a proof gallon contains one half its volume of nearly pure alcohol at 60 degrees Fahrenheit. The term gallon also occurs under Liquid Measure, Apothecaries' Fluid Measure, and Cooking Measure in Part Three, and under Avoirdupois Weight in Part Four.

1 IMPERIAL
GALLON

1.20099 (r) gallons

0.984000 (a) ale gallon

0.400000 (e) barn gallon

0.100000 (e) anker

4.5460665 (r) liters

4.5461938 (r) cubic decimeters

The wine gallon is another name that has denoted both the gallon and imperial gallon, but a wine gallon in Barbados equals 0.83767 gallon. The term imperial gallon is also used in Liquid Measure and Apothecaries' Fluid Measure, Part Three.

1 ALE GALLON,   1.22051 (r) gallons
BEER GALLON   1.01624 (r) imperial gallons
                     0.406496 (r) barn gallon
                     4.6200000 (a) liters
                     4.6201294 (a) cubic decimeters

The stoop, a unit of beer measure in the Netherlands, equals 1.32 gallons. The term stoop is also used in Liquid Measure, Part Three.

1 QUARTER KEG   The quarter keg equals ¼ keg and about 2.5 gallons or less of ale or beer.

1 BARN GALLON   9.61552 (r) bottles (wine)
                     3.00248 (r) gallons
                     2.50000 (e) imperial gallons
                     2.46000 (a) ale gallons
                     0.250000 (e) anker
                     11.3651663 (r) liters
                     11.3654845 (r) cubic decimeters

The barn gallon is used in the measure of wine.

1 CANTARO   3.57200 (a) gallons
                     0.838498 (r) arroba
                     13.5211108 (r) liters
                     13.5214902 (r) cubic decimeters

The cantaro is a wine unit of Spain.

1 ARROBA   4.26000 (a) gallons
                     1.19261 (r) cantaros
                     16.1254120 (r) liters
                     16.1258644 (r) cubic decimeters

The arroba is a Spanish unit used in the measure of wine. The term arroba also occurs under Weight and Mass Units of Spain in Part Four and under Capacity and Volume Units of Spain in Part Three.

1 HALF KEG   The half keg equals ½ keg and about 5.0 gallons of ale or beer.

1 FIRKIN

72.00000 (e) pints (liquid)

36.00000 (e) quarts (liquid)

9.00000 (e) gallons

7.49336 (r) imperial gallons

7.37397 (r) ale gallons

3.77607 (r) yards

34.0677504 (r) liters

34.0687044 (r) cubic decimeters

The firkin is used in the measure of ale and beer, but as a British unit the firkin is often used in the measure of butter. The term firkin also occurs under Avoirdupois Weight in Part Four.

1 KEG

The keg is a small cask used to contain ale or beer. The unit equals 10 gallons or less, 2.0 half kegs, and 4.0 quarter kegs.

1 ANKER

96.07596 (r) pints (liquid)

60.00000 (e) bottles (wine)

48.03798 (r) quarts (liquid)

24.01899 (r) magnums

15.01187 (r) jeroboams

12.00990 (r) gallons

10.00000 (e) imperial gallons

4.00000 (e) barn gallons

0.667194 (r) runlet

0.634921 (r) octave

0.400317 (r) aum

0.381254 (r) barrel (wine)

45.4606652 (r) liters

45.4619380 (r) cubic decimeters

The anker is a British unit used in the measure of wine. The Dutch anker, also used for wine, equals ¼ aam and 9 to 10.5 wine gallons. Among other similar wine units, the antal of Hungary equals 13.4 gallons and 51.0 liters; the orna of Trieste equals 15.1 gallons and 57 liters; the barile of Rome equals 15.41 gallons; and the barile or barril of Greece equals 19.6 gallons.

1 RUNLET,	144.00000 (e) pints (liquid)
RUNDLET,	89.92583 (r) bottles (wine)
RUDLET	72.00000 (e) quarts (liquid)
	36.00000 (e) magnums
	22.50000 (e) jeroboams
	18.00000 (e) gallons
	2.00000 (e) firkins
	1.49876 (r) ankers
	0.951590 (r) octave
	0.600000 (e) aum
	0.571428 (r) barrel (wine)
	0.500000 (e) barrel (ale and beer)
	68.1355008 (r) liters
	68.1374088 (r) cubic decimeters

The unit is used in the measure of wine. It is also known as a kilderkin, a unit also used in the measure of ale and beer.

1 OCTAVE	75.66237 (r) quarts (liquid)
	37.83119 (r) magnums
	25.64449 (r) jeroboams
	18.91559 (r) gallons
	15.75000 (e) imperial gallons
	6.30000 (e) barn gallons
	1.57500 (e) ankers
	1.05087 (r) runlets
	0.630520 (r) aum
	0.600495 (r) barrel (wine)
	0.125000 (e) butt (wine)
	71.6005477 (r) liters
	71.6025524 (r) cubic decimeters

The octave is used in the measure of wine. The term octave also occurs under Sound Measure and Units Used in Music in Part Six.

1 BRITISH	126.00000 (e) bottles (wine)
BOTTLE	100.88316 (r) quarts (liquid)

1 British bottle	25.22079 (r) gallons
(continued)	21.00000 (e) imperial gallons
	2.10000 (e) ankers
	95.4673969 (r) liters
	95.4700698 (r) cubic decimeters

1 AUM
- 120.00000 (e) quarts (liquid)
- 60.00000 (e) magnums
- 37.50000 (e) jeroboams
- 30.00000 (e) gallons
- 2.49794 (r) ankers
- 1.66667 (r) runlets
- 1.58599 (r) octaves
- 0.952381 (r) barrel (wine)
- 0.714286 (r) tierce
- 113.5591680 (r) liters
- 113.5623480 (r) cubic decimeters

The aum is used in the measure of wine.

1 BARREL (WINE)
- 126.00000 (e) quarts (liquid)
- 63.00000 (e) magnums
- 39.37500 (e) jeroboams
- 31.50000 (e) gallons
- 2.62275 (r) ankers
- 1.75000 (e) runlets
- 1.66524 (r) octaves
- 1.05000 (e) aums
- 0.875000 (e) barrel (ale and beer)
- 0.750000 (e) tierce
- 0.416309 (r) British hogshead (wine)
- 119.2371264 (r) liters
- 119.2404654 (r) cubic decimeters

The term barrel also occurs under Liquid Measure and Dry Measure in Part Three and under Avoirdupois Weight in Part Four.

1 BARREL (ALE AND BEER)
- 144.00000 (e) quarts (liquid)
- 36.00000 (e) gallons

4.00000 (e) firkins

2.00000 (e) kilderkins

1.14286 (r) barrels (**wine**)

0.555064 (r) British hogshead (ale and beer)

0.500000 (e) puncheon (ale and beer)

136.2710016 (r) liters

136.2748176 (r) cubic decimeters

The English beer barrel equals 43.23 gallons and 163.65 liters. The term barrel also occurs under Liquid Measure and Dry Measure in Part Three and under Avoirdupois Weight in Part Four.

1 TIERCE

168.00000 (e) quarts (liquid)

84.00000 (e) magnums

52.50000 (e) jeroboams

42.00000 (e) gallons

3.49712 (r) ankers

2.33333 (r) runlets

2.22039 (r) octaves

1.40000 (e) aums

1.33333 (r) barrels (wine)

0.555096 (r) British hogshead (wine)

0.500000 (e) puncheon (wine)

158.9828352 (r) liters

158.9872872 (r) cubic decimeters

The tierce is used in the measure of wine. The term tierce also occurs under Avoirdupois weight in Part Four.

The old Dutch and German aam, also used for wine, equaled 4.0 (e) Dutch ankers and 36 to 42 wine gallons.

1 BRITISH
HOGSHEAD
(ALE AND BEER)

259.41384 (r) quarts (liquid)

64.85346 (r) gallons

54.00000 (e) imperial gallons

7.20594 (r) firkins

3.60297 (r) kilderkins

1.80149 (r) barrels (ale and beer)

0.900743 (r) puncheon (ale and beer)

1 British *hogshead* *(ale and beer)* *(continued)*	0.857143 (r) British hogshead (wine) 0.500000 (e) butt (ale and beer) 0.250000 (e) tun (ale and beer) 245.4908467 (r) liters 245.4977204 (r) cubic decimeters

The term hogshead also occurs under Liquid Measure in Part Three and under Avoirdupois Weight in Part Four.

1 PUNCHEON **(ALE AND BEER)**	288.00000 (e) quarts (liquid) 72.00000 (e) gallons 8.00000 (e) firkins 4.00000 (e) kilderkins 2.00000 (e) barrels (ale and beer) 1.11019 (r) British hogsheads (ale and beer) 0.857143 (r) puncheon (wine) 0.555096 (r) butt (ale and beer) 0.277548 (r) tun (ale and beer) 272.5420032 (r) liters 272.5496352 (r) cubic decimeters

1 BRITISH **HOGSHEAD** **(WINE)**	302.64948 (r) quarts (liquid) 151.32474 (r) magnums 94.57796 (r) jeroboams 75.66237 (r) gallons 63.00000 (e) imperial gallons 6.30000 (e) ankers 4.20346 (r) runlets 4.00000 (e) octaves 2.52208 (r) aums 2.40198 (r) barrels (wine) 1.80196 (r) tierces 1.16667 (r) British hogsheads (ale and beer) 0.900981 (r) puncheon (wine) 0.500000 (e) butt (wine) 0.250000 (e) tun (wine) 286.4021908 (r) liters 286.4102096 (r) cubic decimeters

The term hogshead also occurs under Liquid Measure in Part Three and under Avoirdupois Weight in Part Four.

1 PUNCHEON
(WINE)

336.00000 (e) quarts (liquid)
168.00000 (e) magnums
105.00000 (e) jeroboams
84.00000 (e) gallons
69.93806 (r) imperial gallons
6.99381 (r) ankers
4.66667 (r) runlets
4.44051 (r) octaves
2.80000 (e) aums
2.66667 (r) barrels (wine)
2.00000 (e) tierces
1.16667 (r) puncheons (ale and beer)
1.11013 (r) British hogsheads (wine)
0.555063 (r) butt (wine)
0.277532 (r) tun (wine)
317.9656704 (r) liters
317.9745744 (r) cubic decimeters

1 BUTT, PIPE
(ALE AND BEER)

129.70692 (r) gallons
108.00000 (e) imperial gallons
14.41188 (r) firkins
7.20594 (r) kilderkins
3.60297 (r) barrels (ale and beer)
2.00000 (e) British hogsheads (ale and beer)
1.80149 (r) puncheons (ale and beer)
0.857142 (r) butt (wine)
0.500000 (e) tun (ale and beer)
490.9816934 (r) liters
490.9954408 (r) cubic decimeters

The butt, now a British unit, was first used by the ancient Greeks; the word *butt* possibly gave rise to the word *bottle*.

1 BUTT, PIPE
(WINE)

151.32474 (r) gallons
126.00000 (e) imperial gallons

1 Butt, pipe	12.60000 (e) ankers
(wine)	8.41323 (r) runlets
(continued)	8.00000 (e) octaves
	5.04416 (r) aums
	4.80396 (r) bottles (wine)
	3.60297 (r) tierces
	2.00000 (e) British hogsheads (wine)
	1.80149 (r) puncheons (wine)
	1.16667 (r) butts (ale and beer)
	0.500000 (e) tun (wine)
	572.8043816 (r) liters
	572.8204192 (r) cubic decimeters

1 TUN	259.41384 (r) gallons
(ALE AND BEER)	216.00000 (e) imperial gallons
	28.82376 (r) firkins
	14.41188 (r) kilderkins
	7.20594 (r) barrels (ale and beer)
	4.00000 (e) British hogsheads (ale and beer)
	3.60297 (r) puncheons (ale and beer)
	2.00000 (e) butts (ale and beer)
	0.857142 (r) tun (wine)
	981.9633868 (r) liters
	981.9908816 (r) cubic decimeters

1 TUN (WINE)	302.64948 (r) gallons
	252.00000 (e) imperial gallons
	100.80000 (e) barn gallons
	25.20000 (e) ankers
	16.81386 (r) runlets
	16.00000 (e) octaves
	12.00000 (e) British bottles
	9.60792 (r) barrels (wine)
	7.20594 (r) tierces
	4.00000 (e) British hogsheads (wine)
	3.60297 (r) puncheons (wine)
	2.00000 (e) butts (wine)
	1.16667 (r) tuns (ale and beer)

1,145.6087632 (r) liters
1,145.6408384 (r) cubic decimeters

Comparative capacities and volumes of cubic, liquid, dry, apothecaries' fluid, cooking, and spirits measures

The following table is presented to help the reader visualize the actual sizes of the various capacity and volume units mentioned in the preceding sections of Part Three. The figures given apply to the sides of cubes and the radii of spheres, each cube and sphere having a capacity or volume equivalent to that of the unit it illustrates.

To identify the section in which a particular unit appears, the following code is used:

(**C**) Cubic Measure
(**L**) Liquid Measure
(**D**) Dry Measure
(**F**) Apothecaries' Fluid Measure
(**K**) Cooking Measure
(**S**) Spirits Measure

	Side		Radius	
	Inches	*Centimeters*	*Inches*	*Centimeters*
Cubic inch (C)	1.000(e)	2.540(r)	0.620(r)	1.576(r)
Cup (K)	1.130(r)	2.870(r)	0.701(r)	1.781(r)
British fluid ounce (F)	1.202(r)	3.052(r)	0.745(r)	1.893(r)
Fluid ounce (F)	1.218(r)	3.092(r)	0.755(r)	1.919(r)
Wine glass (F)	1.533(r)	3.894(r)	0.951(r)	2.416(r)
Teacup (F)	1.760(r)	4.460(r)	1.092(r)	2.767(r)
English breakfast cup (F)	1.866(r)	4.740(r)	1.158(r)	2.940(r)

	Side		Radius	
	Inches	*Centimeters*	*Inches*	*Centimeters*
Gill (L)	1.933(r)	4.909(r)	1.199(r)	3.046(r)
Imperial gill (L)	2.058(r)	5.227(r)	1.277(r)	3.243(r)
Mutchkin (L)	2.963(r)	7.526(r)	1.838(r)	4.670(r)
Pint (L, K, S),				
fluid pint (F)	3.067(r)	7.790(r)	1.903(r)	4.833(r)
Pint (D)	3.226(r)	8.195(r)	2.002(r)	5.084(r)
British pint (F)	3.241(r)	8.232(r)	2.011(r)	5.108(r)
Imperial pint (L)	3.260(r)	8.281(r)	2.023(r)	5.138(r)
Fifth (S)	3.588(r)	9.114(r)	2.226(r)	5.654(r)
Bottle (wine) (S)	3.589(r)	9.116(r)	2.227(r)	5.656(r)
Quart (L, K, S)	3.866(r)	9.819(r)	2.398(r)	6.092(r)

	Side		Radius	
	Feet	*Decimeters*	*Feet*	*Decimeters*
Imperial quart (L)	0.330(r)	1.005(r)	0.205(r)	0.624(r)
Quart (D)	0.339(r)	1.033(r)	0.210(r)	0.641(r)
Pottle (L)	0.405(r)	1.236(r)	0.252(r)	0.767(r)
Magnum (S)	0.406(r)	1.237(r)	0.252(r)	0.767(r)
Pottle (D)	0.427(r)	1.301(r)	0.265(r)	0.807(r)
Board foot (C)	0.437(r)	1.332(r)	0.271(r)	0.826(r)
Yard (S)	0.437(r)	1.333(r)	0.271(r)	0.827(r)
Jeroboam (S)	0.475(r)	1.447(r)	0.295(r)	0.955(r)
Gallon (L, F, K, S)	0.511(r)	1.558(r)	0.317(r)	0.967(r)
Imperial gallon				
(L, F, S)	0.544(r)	1.658(r)	0.338(r)	1.029(r)
Ale gallon (S)	0.547(r)	1.666(r)	0.339(r)	1.033(r)
Peck (D, K)	0.678(r)	2.066(r)	0.420(r)	1.282(r)
Barn gallon (S)	0.738(r)	2.248(r)	0.458(r)	1.395(r)
Bucket (D)	0.853(r)	2.602(r)	0.481(r)	1.614(r)
Pin (L)	0.897(r)	2.735(r)	0.506(r)	1.697(r)
Cubic foot (C)	1.000(e)	3.048(r)	0.620(r)	1.891(r)
Firkin (S)	1.064(r)	3.244(r)	0.660(r)	2.012(r)
Bushel (D, K)	1.076(r)	3.280(r)	0.667(r)	2.035(r)

	Side		Radius	
	Feet	*Decimeters*	*Feet*	*Decimeters*
Firlot (D)	1.087(r)	3.312(r)	0.674(r)	2.056(r)
British bushel (D)	1.088(r)	3.317(r)	0.675(r)	2.058(r)
Anker (S)	1.171(r)	3.569(r)	0.726(r)	2.214(r)
Runlet, kilderkin (S)	1.340(r)	4.084(r)	0.831(r)	2.534(r)
Strike (D)	1.355(r)	4.131(r)	0.840(r)	2.563(r)
Octave (S)	1.362(r)	4.153(r)	0.844(r)	2.576(r)
British bottle (S)	1.500(r)	4.570(r)	0.930(r)	2.836(r)
Bag (D)	1.551(r)	4.728(r)	0.962(r)	2.933(r)
Aum (S)	1.582(r)	4.822(r)	0.982(r)	2.992(r)
Barrel (D)	1.598(r)	4.872(r)	0.992(r)	3.023(r)
Barrel (L, S)	1.615(r)	4.922(r)	1.002(r)	3.054(r)
Windle (D)	1.666(r)	4.978(r)	1.034(r)	3.088(r)
Barrel (ale and beer) (S)	1.688(r)	5.146(r)	1.047(r)	3.193(r)
Bulk barrel (C)	1.710(r)	5.212(r)	1.061(r)	3.234(r)
Boll, coombe (D)	1.726(r)	5.261(r)	1.071(r)	3.264(r)
Tierce (S)	1.777(r)	5.417(r)	1.103(r)	3.361(r)
Cran (L)	1.818(r)	5.545(r)	1.129(r)	3.440(r)
Hogshead (L)	2.035(r)	6.202(r)	1.262(r)	3.847(r)
British hogshead (ale and beer)	2.054(r)	6.262(r)	1.275(r)	3.885(r)
Puncheon (ale and beer) (S)	2.127(r)	6.484(r)	1.322(r)	4.022(r)
Quarter (D)	2.151(r)	6.556(r)	1.335(r)	4.068(r)
British hogshead (wine) (S)	2.163(r)	6.592(r)	1.342(r)	4.090(r)
Puncheon (wine) (S)	2.239(r)	6.825(r)	1.389(r)	4.235(r)
Cord foot (C)	2.520(r)	7.681(r)	1.563(r)	4.765(r)
Standard (C)	2.555(r)	7.788(r)	1.584(r)	4.828(r)
Butt (ale and beer) (S)	2.588(r)	7.889(r)	1.606(r)	4.894(r)
Butt (wine) (S)	2.725(r)	8.305(r)	1.690(r)	5.152(r)
Perch (C)	2.914(r)	8.872(r)	1.808(r)	5.504(r)

	Side		Radius	
	Yards	*Meters*	*Yards*	*Meters*
Cubic yard (C)	1.000(e)	0.914(r)	0.620(r)	0.567(r)
Tun (ale and beer) (S)	1.087(r)	0.994(r)	0.674(r)	0.616(r)
Displacement ton (C)	1.090(r)	0.997(r)	0.677(r)	0.618(r)
Water ton (L)	1.100(r)	1.006(r)	0.683(r)	0.624(r)
Marine ton (C)	1.140(r)	1.042(r)	0.707(r)	0.646(r)
Tun (wine) (S)	1.144(r)	1.046(r)	0.710(r)	0.649(r)
Timber load (C)	1.229(r)	1.124(r)	0.762(r)	0.697(r)
Load (D)	1.239(r)	1.133(r)	0.769(r)	0.703(r)
Register ton (C)	1.547(r)	1.415(r)	0.960(r)	0.878(r)
Last (D)	1.562(r)	1.428(r)	0.969(r)	0.886(r)
Stack (C)	1.588(r)	1.452(r)	0.985(r)	0.901(r)
Tank (L)	1.605(r)	1.468(r)	0.996(r)	0.910(r)
Cord (C)	1.680(r)	1.536(r)	1.042(r)	0.953(r)
British standard (C)	1.828(r)	1.672(r)	1.134(r)	1.037(r)
Keel (C)	2.349(r)	2.148(r)	1.458(r)	1.333(r)
Load (C)	3.095(r)	2.830(r)	1.920(r)	1.756(r)
Shipload (C)	6.376(r)	5.831(r)	3.956(r)	3.618(r)
Acre-foot (C)	11.728(r)	8.937(r)	7.276(r)	5.545(r)

Intermeasure comparison of capacity and volume units

The following material shows the relation between various units of capacity and volume. To identify the section in which a particular unit appears, the following code is used:

(**C**) Cubic Measure

(**L**) Liquid Measure

(**D**) Dry Measure

(**F**) Apothecaries' Fluid Measure

(**K**) Cooking Measure

(**S**) Spirits Measure

1 CUBIC **INCH (C)**	1,384.19221 (r) fluid scruples (F) 0.738778 (r) pony (S) 0.014880 (r) quart (D)
1 PONY (S)	1.35351 (r) cubic inches (C) 0.781344 (r) British fluid ounce (F)
1 BRITISH FLUID **OUNCE (F)**	1.27985 (r) ponies (S) 0.639921 (r) jigger (S)
1 SHOT (S), **FLUID** **OUNCE (F)**	1.80468 (r) cubic inches (C) 0.250000 (e) gill (L)
1 JIGGER (S)	2.70703 (r) cubic inches (C) 1.56234 (r) British fluid ounces (F)
1 ENGLISH **BREAKFAST** **CUP (F)**	6.49685 (r) cubic inches (C) 0.749389 (r) imperial gill (L)
1 GILL (L)	7.21873 (r) cubic inches (C) 0.013427 (r) peck (D, K)
1 PINT (L, K, S), **FLUID PINT (F)**	28.87494 (r) cubic inches (C) 0.859347 (r) pint (D)
1 PINT (D)	33.60118 (r) cubic inches (C) 1.16368 (r) pints (L, K, S), fluid pints (F)
1 FIFTH (S)	46.20018 (r) cubic inches (C) 5.32899 (r) imperial gill (L)
1 QUART **(L, K, S)**	57.74987 (r) cubic inches (C) 0.859343 (r) quart (D)
1 QUART (D)	66.79960 (r) cubic inches (C) 1.16368 (r) quarts (L, K, S)

| 1 IMPERIAL QUART (L) | 38.43089 (r) fluid ounces (F) |
| | 0.040135 (r) cubic foot (C) |

| 1 BOARD FOOT (C) | 4.28568 (r) pints (D) |
| | 2.49352 (r) quarts (L) |

| 1 GALLON (L, F, K, S) | 3.43737 (r) quarts (D) |
| | 0.133674 (r) cubic foot (C) |

| 1 IMPERIAL GALLON (L, F, S) | 19.21568 (r) cups (K) |
| | 0.160540 (r) cubic foot (C) |

| 1 PECK (D, K) | 537.61559 (r) cubic inches (C) |
| | 7.75155 (r) imperial quarts (L) |

| 1 CUBIC FOOT (C) | 25.71449 (r) quarts (D) |
| | 7.48085 (r) gallons (L, F, K, S) |

| 1 BUSHEL (D, K) | 62.02439 (r) imperial pints (L) |
| | 1.24585 (r) cubic feet (C) |

| 1 BRITISH BUSHEL (D) | 38.43155 (r) quarts (L) |
| | 1.28433 (r) cubic feet (C) |

| 1 BARREL (D) | 0.969733 (r) barrel (L) |
| | 0.816660 (r) bulk barrel (C) |

| 1 BULK BARREL (C) | 1.22450 (r) barrels (D) |
| | 1.18739 (r) barrels (L) |

| 1 HOGSHEAD (L) | 2.06242 (r) barrels (D) |
| | 1.68429 (r) bulk barrels (C) |

| 1 CORD FOOT (C) | 111.76380 (r) gallons (L) |
| | 1.60711 (r) quarters (D) |

| 1 PERCH (C) | 2.48600 (r) quarters (D) |
| | 2.93891 (r) hogsheads (L) |

1 CUBIC **YARD (C)**	694.29120 (r) quarts (D) 3.20608 (r) hogsheads (L) 0.525551 (r) load (D)
1 DISPLACEMENT **TON (C)**	8.31207 (r) barrels (L) 0.681270 (r) load (D)
1 MARINE **TON (C)**	4.74975 (r) hogsheads (L) 0.778596 (r) load (D)
1 LOAD (D)	6.10042 (r) hogsheads (L) 1.02749 (r) timber load (C)
1 TANK (L)	1.58730 (r) tuns (wine) (S) 0.009174 (r) shipload (C)
1 REGISTER **TON (C)**	11.87438 (r) hogsheads (L) 0.973243 (r) last (D)
1 CORD (C)	30.39841 (r) barrels (L) 1.24575 (r) lasts (D)
1 LOAD (C)	7.78594 (r) lasts (D) 1.24579 (r) tanks (L)

Capacity and volume units of ancient Arabia

The following units may prove useful to the archeologist and to students of the arts and history.

| **1 CAPHITE,**
KILADJA, KIST | 1.45200 (a) quarts (liquid)
0.500000 (e) saa
0.333333 (r) makuk
0.166667 (r) ferk
0.083333 (r) woibe |

1 Caphite,	1.3753908 (a) liters
kiladja, kist	1.3754293 (a) cubic decimeters
(continued)	

1 SAA

2.90700 (a) quarts (liquid)
2.00000 (e) caphites
0.726700 (a) gallon
0.666667 (r) makuk
0.333333 (r) ferk
0.166667 (r) woibe
0.083333 (r) cafiz
2.7507816 (a) liters
2.7508586 (a) cubic decimeters

1 MAKUK

3.75000 (a) quarts (dry)
3.00000 (e) caphites
1.50000 (e) saas
1.09000 (a) gallons
0.500000 (e) ferk
0.250000 (e) woibe
0.125000 (e) cafiz
4.1261724 (a) liters
4.1262879 (a) cubic decimeters
The makuk was also used in Assyria, Chaldea, and Persia.

1 FERK

6.00000 (e) caphites
3.00000 (e) saas
2.18000 (a) gallons
2.00000 (e) makuks
0.500000 (e) woibe
0.250000 (e) cafiz
0.234000 (a) bushel
0.125000 (e) artaba
8.2523447 (a) liters
8.2525758 (a) cubic decimeters

1 WOIBE

12.00000 (e) caphites
4.36000 (a) gallons

4.00000 (e) makuks
0.500000 (e) cafiz
0.468500 (a) bushel
16.5046895 (a) liters
16.5051517 (a) cubic decimeters

1 CAFIZ   24.00000 (e) caphites
8.72000 (a) gallons
8.00000 (e) makuks
0.937000 (a) bushel
0.500000 (e) artaba
33.0093790 (a) liters
33.0103033 (a) cubic decimeters

1 ARTABA   24.00000 (e) saas
17.44000 (a) gallons
2.00000 (e) cafices
1.87400 (a) bushels
0.250000 (e) den
66.0187579 (a) liters
66.0206066 (a) cubic decimeters

The artaba was also used in Assyria, Chaldea, and
Persia.

1 DEN, GARIBA   69.77000 (a) gallons
64.00000 (e) makuks
8.00000 (e) cafices
7.49600 (a) bushels
264.0750316 (a) liters
264.0824264 (a) cubic decimeters

The unit was also used in Assyria, Chaldea, and
Persia.

Capacity and volume units of ancient Greece

The following units may prove useful to the archeologist and to students of the arts and history.

1 OXYBAPHON
0.214836 (r) cotula
0.146718 (r) cyathos
0.140000 (a) pint (liquid)
0.0662430 (a) liter
0.0662450 (a) cubic decimeter

1 COTULA,
HEMINA,
KOTYLE
4.65471 (r) oxybapha
0.682927 (r) cyathos
0.285716 (r) choenix
0.280000 (a) quart (dry)
0.142857 (r) maris
0.3083419 (a) liter
0.3083505 (a) cubic decimeter

1 CYATHOS
6.81579 (r) oxybapha
1.46429 (r) cotulai
0.820000 (a) pint (dry)
0.418367 (r) choenix
0.4515006 (a) liter
0.4515132 (a) cubic decimeter

1 XESTES
2.00000 (e) cotulai
0.6166838 (a) liter
0.6167010 (a) cubic decimeter

1 CHOENIX
16.29147 (r) oxybapha
3.50000 (e) cotulai
2.39024 (r) cyathoi
0.500000 (e) maris
0.250000 (e) hemiekton
0.125000 (e) hekteus
0.122500 (a) peck
1.0791967 (a) liters
1.0792268 (a) cubic decimeters

1 MARIS
32.58198 (r) oxybapha
7.00000 (e) cotulai
4.78048 (r) cyathoi

2.00000 (e) choenices
0.500000 (e) hemiekton
0.250000 (e) hekteus
0.245000 (a) peck
2.1583936 (a) liters
2.1584536 (a) cubic decimeters

1 KHOUS   3.00000 (e) choenices
3.2375901 (a) liters
3.2376804 (a) cubic decimeters

1 HEMIEKTON   65.16588 (r) oxybapha
14.00000 (e) cotulai
9.65000 (r) cyathoi
4.00000 (e) choenices
3.95650 (a) quarts (dry)
2.00000 (e) mareis
0.500000 (e) hekteus
0.490000 (a) peck
4.3167872 (a) liters
4.3169072 (a) cubic decimeters

1 HEKTEUS   130.33176 (r) oxybapha
28.00000 (e) cotulai
19.30000 (r) cyathoi
8.00000 (e) choenices
4.00000 (e) mareis
2.00000 (e) hemiekta
0.980000 (a) peck
8.6335744 (a) liters
8.6338144 (a) cubic decimeters

The hekteus was also known as a chos, a liquid measure equaling 3.4 (liquid) quarts.

1 CADOS   574.28437 (r) oxybapha
123.37667 (r) cotulai
84.25782 (r) cyathoi

1 Cados
(continued)

35.25069 (r) choenices
17.62535 (r) mareis
10.05000 (a) gallons
8.81267 (r) hemiekta
4.40634 (r) hekteis
0.975444 (r) metretes
0.755223 (r) medimnos
38.0423213 (a) liters
38.0433566 (a) cubic decimeters

The cados also equaled 1.5 amphorae; see Capacity and Volume Units of Ancient Rome in Part Three.

**1 METRETES,
AMPHURA**

588.69122 (r) oxybapha
126.47600 (r) cotulai
86.37371 (r) cyathoi
36.13600 (r) choenices
18.06800 (r) mareis
10.30000 (a) gallons
9.03450 (r) hemiekta
4.51725 (r) hekteis
1.02517 (r) cadoi
0.752875 (r) medimnos
39.0000000 (a) liters
39.0010920 (a) cubic decimeters

The metretes is also considered to have equaled 10.4 gallons. The term amphura also occurs under Capacity and Volume Units of Ancient Rome in Part Three.

**1 MEDIMNOS,
MEDIMNUS**

781.96757 (r) oxybapha
168.00000 (e) cotulai
114.73162 (r) cyathoi
48.00000 (e) choenices
24.00000 (e) mareis
12.00000 (e) hemiekta
6.00000 (e) hekteis
5.88000 (a) pecks
1.47000 (a) bushels

1.36168 (r) cadoi
1.32824 (r) metretai
51.8014416 (a) liters
51.8028864 (a) cubic decimeters

Capacity and volume units of ancient India

The following units may prove useful to the archeologist and to students of the arts and history.

1 MUSHTI
0.250000 (e) cudava
0.136000 (a) pint (liquid)
0.117200 (a) pint (dry)
0.0609636 (a) liter
0.0609653 (a) cubic decimeter

1 CUDAVA
4.00000 (e) mushtis
0.544000 (a) pint (liquid)
0.468750 (a) pint (dry)
0.250000 (e) prastha
0.2438544 (a) liter
0.2438612 (a) cubic decimeter

1 PRASTHA
16.00000 (e) mushtis
4.00000 (e) cudavas
2.17000 (a) pints (liquid)
1.87500 (a) pints (dry)
0.9754174 (a) liter
0.9754447 (a) cubic decimeter

1 DRONA
256.00000 (e) mushtis
64.00000 (e) cudavas
34.80000 (a) pints (liquid)
30.00000 (a) pints (dry)
16.00000 (e) prasthas

1 Drona	15.6066782 (a) liters
(continued)	15.6071153 (a) cubic decimeters

1 CUMBHA   696.00000 (a) pints (liquid)
600.00000 (a) pints (dry)
320.00000 (e) prasthas
20.00000 (e) dronas
312.1335648 (a) liters
312.1423058 (a) cubic decimeters

Capacity and volume units of ancient Israel

The following units may prove useful to the archeologist and to students of the arts and history.

1 CAB   2.00000 (a) quarts (dry)
2.2024420 (a) liters
2.2025036 (a) cubic decimeters

1 OMER   1.80000 (r) cabs
0.450000 (a) peck
3.9643954 (a) liters
3.9645065 (a) cubic decimeters

1 HEKAT   291.00000 (a) cubic inches
4.7686356 (a) liters
4.7686752 (a) cubic decimeters

1 HIN   1.67970 (r) gallons
1.33333 (r) hekats
6.3581649 (a) liters
6.3582177 (a) cubic decimeters

1 BATH, BU   2,250.00000 (a) cubic inches
36.8708940 (a) liters
36.8712000 (a) cubic decimeters

---

1 EPHAH

10.00000 (a) omers
1.07521 (a) bu
39.6439540 (a) liters
39.6450648 (a) cubic decimeters

The ephah is also considered to have equaled 1.1 bu.

1 KOR (LIQUID)

97.40522 (a) gallons
10.00000 (r) baths
368.7089401 (a) liters
368.7120000 (a) cubic decimeters

The liquid kor is also considered to have equaled 370 liters and 100 gallons.

1 KOR (DRY),
HOMER

10.00000 (r) ephahs
396.4395400 (a) liters
396.4506480 (a) cubic decimeters

The unit is also considered to have equaled 400 liters.

Capacity and volume units of
ancient Rome

The following units may prove useful to the archeologist and to students of the arts and history.

1 ACETABULUM

0.500000 (e) quartarius
0.140000 (a) pint (liquid)
0.120000 (a) pint (dry)
0.0660733 (a) liter
0.0660752 (a) cubic decimeter

1 QUARTARIUS

2.00000 (e) acetabula
0.500000 (e) hemina
0.250000 (e) sextarius
0.240000 (a) pint (dry)
0.140000 (a) quart (liquid)

| 1 *Quartarius* (continued) | 0.1321465 (a) liter |
| | 0.1321503 (a) cubic decimeter |

The quartarius is also considered to have equaled 0.145 liter.

1 HEMINA
2.00000 (e) quartarii
0.571428 (r) cyathus
0.560000 (a) pint (liquid)
0.500000 (e) sextarius
0.480000 (a) pint (dry)
0.2642930 (a) liter
0.2643006 (a) cubic decimeter

1 CYATHUS
7.00000 (e) acetabula
1.75000 (e) heminae
0.980000 (a) pint (liquid)
0.875000 (e) sextarius
0.850000 (a) pint (dry)
0.4625128 (a) liter
0.4625261 (a) cubic decimeter

1 SEXTARIUS
2.00000 (e) heminae
1.14286 (r) cyathi
1.12000 (a) pints (liquid)
0.960000 (a) pint (dry)
0.5285861 (a) liter
0.5286012 (a) cubic decimeter

The sextarius is also considered to have equaled 0.58 liter.

1 CONGIUS
12.00000 (e) heminae
6.85714 (r) cyathi
6.72000 (a) pints (liquid)
6.00000 (e) sextarii
5.76000 (a) pints (dry)
0.375000 (e) modius
3.1715163 (a) liters
3.1716072 (a) cubic decimeters

1 MODIUS 32.00000 (e) heminae
18.28571 (r) cyathi
17.92000 (a) pints (liquid)
16.00000 (e) sextarii
15.36000 (a) pints (dry)
2.66667 (r) congii
0.666667 (r) urna
8.4573768 (a) liters
8.4576192 (a) cubic decimeters

1 URNA 27.42857 (r) cyathi
26.88000 (a) pints (liquid)
24.00000 (e) sextarii
23.04000 (a) pints (dry)
4.00000 (e) congii
3.37000 (a) gallons
1.50000 (e) modii
0.500000 (e) amphura
12.6860653 (a) liters
12.6864288 (a) cubic decimeters

The urna is also considered to have equaled 3.42 gallons and 12.9457452 (r) liters.

1 AMPHURA 54.85714 (r) cyathi
53.76000 (a) pints (liquid)
46.08000 (a) pints (dry)
8.00000 (e) congii
6.74000 (a) gallons
3.00000 (e) modii
2.00000 (e) urnae
25.3721305 (a) liters
25.3728576 (a) cubic decimeters

1 CULEUS, 1,920.00000 (e) heminae
DOLIUM 1,097.14286 (r) cyathi
1,075.00000 (a) pints (liquid)
921.60000 (a) pints (dry)

1 *Culeus, dolium*	134.80000 (a) gallons
(continued)	20.00000 (e) amphurae
	507.4571520 (a) liters
	507.4426100 (a) cubic decimeters

Capacity and volume units of Argentina

Argentina, as is true of most countries, is currently using the metric system, but the following units persist in many measuring instances.

1 FRASCO
2.50963 (a) quarts (liquid)
2.3749244 (a) liters
2.3749909 (a) cubic decimeters

1 BARIL
20.07700 (a) gallons
8.00000 (e) frascos
18.9993951 (a) liters
18.9999272 (a) cubic decimeters

1 FANEGA
3.89333 (a) bushels
137.1973362 (a) liters
137.2011735 (a) cubic decimeters

1 TONELADA
29.20000 (a) bushels
7.50000 (e) fanegas
1,028.9800214 (a) liters
1,029.0088009 (a) cubic decimeters

1 LASTRE
58.40000 (a) bushels
2.00000 (e) toneladas
2,057.9600429 (a) liters
2,058.0176018 (a) cubic decimeters

Capacity and volume units of Austria

Austria, as is true of most countries, is currently using the metric system, but the following units persist in many measuring instances.

1 BECHER
0.087000 (a) pint (dry)
0.0479031 (a) liter
0.0479045 (a) cubic decimeter

1 PFIFF
0.250000 (e) halbe
0.186916 (a) quart (liquid)
0.1768842 (a) liter
0.1768885 (a) cubic decimeter

1 SEIDEL
2.00000 (e) pfiffs
0.3537684 (a) liter
0.3537770 (a) cubic decimeter

1 FUTTER·
MASSEL
10.00000 (e) bechers
0.870000 (a) pint (dry)
0.4790308 (a) liter
0.4790445 (a) cubic decimeter

The futtermassel is also considered to equal 2.0 (e) bechers.

1 HALBE
4.00000 (e) pfiffs
0.747666 (a) quart (liquid)
0.500000 (e) mass
0.7075367 (a) liter
0.7075540 (a) cubic decimeter

1 MASS
8.00000 (e) pfiffs
2.00000 (e) halbes
1.49533 (a) quarts (liquid)
1.4150734 (a) liters
1.4151080 (a) cubic decimeters

1 MUTH MASSEL   80.00000 (e) bechers
3.48000 (a) quarts (dry)
3.8322460 (a) liters
3.8323562 (a) cubic decimeters

The muth massel is also considered to equal 3.49 dry quarts.

1 ACHTEL   8.00000 (e) futtermassels
6.96000 (a) quarts (dry)
2.00000 (e) muth massels
0.500000 (e) viertel
7.6644920 (a) liters
7.6647124 (a) cubic decimeters

The achtel is also considered to equal 6.98 dry quarts.

1 VIERTEL   13.92000 (a) quarts (dry)
2.00000 (e) achtels
0.435000 (a) bushel
0.250000 (e) metze
15.3289840 (a) liters
15.3294248 (a) cubic decimeters

1 METZE   4.00000 (e) viertels
1.74000 (a) bushels
61.3159360 (a) liters
61.3176992 (a) cubic decimeters

The metze is also considered to equal 1.75 bushels.

1 MUTH   52.20000 (a) bushels
30.00000 (e) metzen
1,839.4780800 (a) liters
1,839.5309760 (a) cubic decimeters

1 DREILING   1,200.00000 (e) mass
448.59900 (a) gallons
1,698.0880800 (a) liters
1,698.1296000 (a) cubic decimeters

Capacity and volume units of Brazil

Brazil, as is true of most countries, is currently using the metric system, but the following units persist in many measuring instances.

1 GARRAFA

0.704000 (a) quart (liquid)
0.6662138 (a) liter
0.6662324 (a) cubic decimeter

1 QUARTO, CUARTA

8.23000 (a) quarts (dry)
1.02875 (a) pecks
0.257189 (a) bushel
9.0630488 (a) liters
9.0633023 (a) cubic decimeters

1 FANGA

16.00000 (e) quartos
4.11500 (a) bushels
145.0087813 (a) liters
145.0128370 (a) cubic decimeters

1 PIPA

126.60000 (a) gallons
0.500000 (e) tonel
479.2196890 (a) liters
479.2331086 (a) cubic decimeters

1 TONEL

253.20000 (a) gallons
2.00000 (e) pipas
958.4393779 (a) liters
958.4662171 (a) cubic decimeters

1 MOIO

61.72500 (a) bushels
15.00000 (e) fangas
2,175.1317192 (a) liters
2,175.1925554 (a) cubic decimeters

The moio is also considered to equal 61.8 bushels.

Capacity and volume units of Cyprus

Cyprus, as is true of most countries, is currently using the metric system, but the following units persist in many measuring instances.

1 OKA, OKE
1.35094 (a) quarts (liquid)
1.2784276 (a) liters
1.2784634 (a) cubic decimeters

1 CASS
5.00000 (a) quarts (liquid)
4.7316320 (a) liters
4.7317645 (a) cubic decimeters

1 KARTOS
5.40375 (a) quarts (liquid)
4.00000 (e) okas
5.1137105 (a) liters
5.1138537 (a) cubic decimeters

1 KOUZA
2.70188 (a) gallons
2.00000 (e) kartos
10.2274211 (a) liters
10.2277075 (a) cubic decimeters

The kouza is also considered to equal 9.0 British imperial quarts and 10.2285494 (r) liters.

1 MEDIMNO
2.13000 (a) bushels
75.0592234 (a) liters
75.0613227 (a) cubic decimeters

1 GOMARI
43.23000 (a) gallons
32.00000 (e) kartos
163.6387369 (a) liters
163.6433192 (a) cubic decimeters

Capacity and volume units of Denmark

Denmark, as is true of most countries, is currently using the metric system, but the following units persist in many measuring instances.

1 PAEGL
0.510000 (a) pint (liquid)
0.250000 (e) pot
0.2413132 (a) liter
0.2413200 (a) cubic decimeter

1 POT
4.00000 (e) paegls
2.04000 (a) pints (liquid)
0.9652529 (a) liter
0.9652801 (a) cubic decimeter

The pot defined above is also used in Norway.

1 ACHTEL
1.97000 (a) quarts (dry)
0.125000 (e) ottingkar
2.1694054 (a) liters
2.1694661 (a) cubic decimeters

1 TONDE (LIQUID)
27.76800 (a) pints (liquid)
13.61177 (a) pots
3.47100 (a) gallons
13.1387957 (a) liters
13.1391637 (a) cubic decimeters

1 OTTINGKAR
8.00000 (e) achtels
0.492500 (a) bushel
17.3552430 (a) liters
17.3557284 (a) cubic decimeters

1 VIERTEL
56.32000 (a) pints (liquid)
27.60784 (a) pots
7.04000 (a) gallons

1 Viertel	26.5485514 (a) liters
(continued)	26.6492977 (a) cubic decimeters

1 FJERDING   0.985000 (a) bushel
0.250000 (e) korntonde
34.7104859 (a) liters
34.7114568 (a) cubic decimeters

The fjerding is also considered to equal 0.988 bushel.

1 OLTONDE   136.11765 (a) pots
34.71000 (a) gallons
10.00000 (e) tonde (liquid)
131.3879574 (a) liters
131.3916366 (a) cubic decimeters

1 KORNTONDE,   64.00000 (e) achtels
TONDE (DRY)   8.00000 (e) ottingkars
4.00000 (e) fjerdings
3.94000 (a) bushels
138.8419437 (a) liters
138.8458269 (a) cubic decimeters

The korntonde is also considered to equal 139.12 liters, and the dry tonde to equal 3.95 bushels. In Norway, the korntonde equals 138.97 liters.

Capacity and volume units of Egypt

Egypt, as is true of most countries, is currently using the metric system, but the following units persist in many measuring instances.

1 TOUMNAH   2.17900 (a) gills
0.272375 (a) quart (liquid)
0.2577557 (a) liter
0.2577629 (a) cubic decimeter

1 ROBHAH

2.00000 (e) toumnahs
0.544750 (a) quart (liquid)
0.500000 (e) nisf keddah
0.5155114 (a) liter
0.5155257 (a) cubic decimeter

The hen, an ancient Egyptian unit, equaled 0.505 liquid quart.

1 NISF KEDDAH

2.00000 (e) robhahs
0.936254 (a) quart (dry)
0.500000 (e) keddah
1.0310228 (a) liters
1.0310515 (a) cubic decimeters

The nisf keddah is also considered to equal 0.963 dry quarts.

1 KEDDAH

8.00000 (e) toumnahs
4.00000 (e) robhahs
2.17900 (a) quarts (liquid)
2.00000 (e) nisf keddahs
1.87251 (a) quarts (dry)
2.0620456 (a) liters
2.0621030 (a) cubic decimeters

1 ROB, ROUB, ROUBOUH

8.71600 (a) quarts (liquid)
4.00000 (e) keddahs
2.17900 (a) gallons
8.2481823 (a) liters
8.2484119 (a) cubic decimeters

1 KILAH

17.43200 (a) quarts (liquid)
14.98007 (a) quarts (dry)
2.00000 (e) robs
0.468127 (a) bushel
16.4963646 (a) liters
16.4968238 (a) cubic decimeters

1 ARTABA   35.20000 (a) quarts (dry)
1.10000 (a) bushels
38.7629792 (a) liters
38.7640634 (a) cubic decimeters

The artaba is an ancient and obsolete unit.

1 APT   59.84000 (a) quarts (dry)
1.87000 (a) bushels
65.8970646 (a) liters
65.8989077 (a) cubic decimeters

The apt is an ancient and obsolete unit.

1 ARDEB   179.80800 (a) quarts (dry)
5.61900 (a) bushels
198.0083456 (a) liters
198.0138837 (a) cubic decimeters

1 DARIBAH   1,438.46400 (a) quarts (dry)
44.95200 (a) bushels
8.00000 (e) ardebs
1,584.0667645 (a) liters
1,584.1110692 (a) cubic decimeters

Capacity and volume units of France

France, as is true of most countries, is currently using the metric system, but the following units persist in many measuring instances.

1 ROQUILLE   0.250000 (e) poisson
0.246000 (a) gill
0.0290995 (a) liter
0.0291004 (a) cubic decimeter

1 POISSON   4.00000 (e) roquilles
0.984000 (a) gill

0.246000 (a) pint (liquid)
0.1163982 (a) liter
0.1164014 (a) cubic decimeter

The poisson is also considered to equal 0.245 liquid pint.

1 CHOPINE

1.06000 (a) pints (liquid)
0.5015530 (a) liter
0.5015671 (a) cubic decimeter

1 PINTE

8.00000 (e) poissons
0.984000 (a) quart (liquid)
0.9311852 (a) liter
0.9312114 (a) cubic decimeter

1 POT

2.00000 (e) pintes
1.96800 (a) quarts (liquid)
1.8623704 (a) liters
1.8624227 (a) cubic decimeters

1 QUARTE

3.43286 (a) quarts (liquid)
2.95000 (a) quarts (dry)
3.2486020 (a) liters
3.2486928 (a) cubic decimeters

1 BOISSEAU

13.73142 (a) quarts (liquid)
11.80000 (a) quarts (dry)
4.00000 (e) quartes
12.9944078 (a) liters
12.9947712 (a) cubic decimeters

1 QUARTAUT

72.00000 (e) pintes
17.71200 (a) gallons
67.0453327 (a) liters
67.0472173 (a) cubic decimeters

1 HEMINE

The hemine, a grain measure formerly used in southern France, is equivalent to about 50.0 liters.

Capacity and volume units of Iceland

Iceland, as is true of most countries, is currently using the metric system, but the following units persist in many measuring instances.

1 POTTAR
1.02111 (a) quarts (liquid)
0.9663034 (a) liter
0.9663304 (a) cubic decimeter

1 KORNSKEPPA
18.38000 (a) quarts (liquid)
18.00000 (a) pottars
17.3934603 (a) liters
17.3939474 (a) cubic decimeters

1 ALMENN-
TURMA
122.53333 (a) quarts (liquid)
120.00000 (e) pottars
30.63333 (a) gallons
115.9564020 (a) liters
115.9596492 (a) cubic decimeters

The almenn-turma is also considered to equal 122.52 liquid quarts.

1 OLTUNNA
138.87096 (a) quarts (liquid)
34.71774 (a) gallons
131.4172556 (a) liters
131.4209357 (a) cubic decimeters

The oltunna is also considered to equal 138.84 liquid quarts.

1 KORNTUNNA
36.76000 (a) gallons
8.00000 (e) kornskeppa
139.1476824 (a) liters
139.1515790 (a) cubic decimeters

The korntunna is also considered to equal 36.75 gallons.

Capacity and volume units of Japan

Japan, as is true of most countries, is currently using the metric system, but the following units persist in many measuring instances.

1 SHO

1.64000 (a) quarts (dry)
1.8060024 (a) liters
1.8060530 (a) cubic decimeters

1 TO

10.00000 (e) sho
2.05000 (a) pecks
18.0600244 (a) liters
18.0605295 (a) cubic decimeters

1 SHAKU

1.10100 (a) cubic feet
31.1761578 (a) liters
31.1770308 (a) cubic decimeters

1 KOKU

190.84349 (a) quarts (liquid)
10.00000 (e) to
180.6052952 (a) liters
180.6002440 (a) cubic decimeters

The koku is also considered to equal 39.68 and 47.655 gallons and 158.72 and 163.808 dry quarts.

1 GO

10.00000 (e) shaku
311.7615778 (a) liters
311.7703076 (a) cubic decimeters

Capacity and volume units of Libya

Libya, as is true of most countries, is currently using the metric system, but the following units persist in many measuring instances.

1 BOZZE   2.83500 (a) quarts (liquid)
                      2.6828353 (a) liters
                      2.6829105 (a) cubic decimeters

1 MISURA   4.36000 (a) gallons
                      16.5039324 (a) liters
                      16.5043946 (a) cubic decimeters

1 MATTARO   6.16000 (a) gallons
                      23.3174825 (a) liters
                      23.3181355 (a) cubic decimeters

1 TEMAN   3.04000 (a) pecks
                      26.7816947 (a) liters
                      26.7824438 (a) cubic decimeters

1 BARILE   24.00000 (e) bozzes
                      17.01000 (a) gallons
                      64.3880483 (a) liters
                      64.3898513 (a) cubic decimeters

Capacity and volume units of Mexico

Mexico, as is true of most countries, is currently
using the metric system, but the following units
persist in many measuring instances.

1 CUARTILLO   0.482000 (a) quart (liquid)
                      0.4561293 (a) liter
                      0.4561421 (a) cubic decimeter

A cuartillo of oil, however, equals 0.535 liquid quart.

1 JARRA   18.00000 (e) cuartillos
                      8.67600 (a) quarts (liquid)
                      8.2103279 (a) liters
                      8.2105578 (a) cubic decimeters

1 CUARTERON   25.0000000 (e) liters
(DRY)   25.0007000 (e) cubic decimeters

1 BARRIL   66.88000 (a) quarts (liquid)
                    63.2903096 (a) liters
                    63.2920820 (a) cubic decimeters

1 FANEGA   82.46400 (a) quarts (dry)
                    2.57700 (a) bushels
                    90.8110885 (a) liters
                    90.8136284 (a) cubic decimeters

1 CARGA   164.92800 (a) quarts (dry)
                    20.00000 (e) fanegas
                    1.8162218 (a) hectoliters
                    1,816.2725687 (a) cubic decimeters

Capacity and volume units of the Netherlands

The Netherlands, or Holland, as is true of most countries, is currently using the metric system, but the following units persist in many measuring instances.

1 VINGERHOED   0.100000 (e) maatje
                    0.10021393 (e) centiliters

The vingerhoed is also considered to equal 1.0 (e) centiliter.

1 MAATJE   0.105895 (a) quart (liquid)
                    0.1002111 (a) liter
                    0.1002139 (a) cubic decimeter

1 KAN, KOP   10.00000 (e) maatjes
                    1.05895 (a) quarts (liquid)
                    0.909522 (a) quart (dry)

1 Kan, kop *(continued)*	1.0021112 (a) liters 1.0021393 (a) cubic decimeters

The kan is used in liquid measure, the kop in dry measure.

1 MINGELEN	1.28333 (a) quarts (liquid) 1.2144491 (a) liters 1.2144831 (a) cubic decimeters

1 SCHEPEL	10.00000 (e) koppen 9.09521 (a) quarts (dry) 1.13690 (a) pecks 10.0211121 (a) liters 10.0213927 (a) cubic decimeters

The schepel is also considered to equal 1.135 pecks.

1 ZAK	90.95215 (a) quarts (dry) 10.00000 (e) schepels 2.84226 (a) bushels 100.2111206 (a) liters 100.2139268 (a) cubic decimeters

The zak is also considered to equal 2.838 bushels.

1 OKSHOOFD	246.40000 (a) quarts (liquid) 192.00000 (e) mingelens 233.1742193 (a) liters 233.1807489 (a) cubic decimeters

1 WISSE	909.52147 (a) quarts (dry) 28.42255 (a) bushels 10.00000 (e) zak 1.31149 (a) cubic yards 1,002.1112057 (a) liters 1,002.1392676 (a) cubic decimeters

Capacity and volume units of Poland

Poland, as is true of most countries, is currently using the metric system, but the following units persist in many measuring instances.

1 KWARTERKA
0.264000 (a) quart (liquid)
0.2498302 (a) liter
0.2498372 (a) cubic decimeter

1 KWARTA
4.00000 (e) kwarterkas
1.05600 (a) quarts (liquid)
0.9993207 (a) liter
0.9993487 (a) cubic decimeter

1 GARNIEC
4.00000 (e) kwartas
1.05600 (a) gallons
3.9972827 (a) liters
3.9972947 (a) cubic decimeters

The garniec is also considered to equal 1.057 gallons.

1 CWIERC
8.44800 (a) gallons
8.00000 (e) garniec
31.9782617 (a) liters
31.9791572 (a) cubic decimeters

1 KORZEC
33.79200 (a) gallons
4.00000 (e) cwierc
3.62987 (a) bushels
127.9130468 (a) liters
127.9166288 (a) cubic decimeters

Capacity and volume units of Portugal

Portugal, as is true of most countries, is currently using the metric system, but the following units persist in many measuring instances.

1 SELAMIN   0.393000 (a) quart (dry)
0.4327799 (a) liter
0.4327920 (a) cubic decimeter

1 MEIO (LIQUID)   0.737167 (a) quart (liquid)
0.6976000 (a) liter
0.6976198 (a) cubic decimeter

1 CANADA   1.47433 (a) quarts (liquid)
0.083333 (r) almude
1.3952004 (a) liters
1.3952396 (a) cubic decimeters

The canada is used mostly in **Lisbon.**

1 OITAVO   4.00000 (e) selamins
1.7311196 (a) liters
1.7311680 (a) cubic decimeters

1 QUARTO   8.00000 (e) selamins
3.14400 (a) quarts (dry)
3.4622388 (a) liters
3.4623357 (a) cubic decimeters

1 MEIO (DRY)   6.28800 (a) quarts (dry)
2.00000 (e) quartos
6.9244777 (a) liters
6.9246713 (a) cubic decimeters

1 ALQUEIRE   12.57600 (a) quarts (dry)
2.00000 (e) meios (dry)
13.8489553 (a) liters
13.8493426 (a) cubic decimeters

1 ALMUDE   24.00000 (e) meios (liquid)
17.69200 (a) quarts (liquid)
4.42300 (a) gallons

16.7424047 (a) liters
16.7428755 (a) cubic decimeters

1 FANGA

50.30400 (a) quarts (dry)
1.57200 (a) bushels
55.3958212 (a) liters
55.3973706 (a) cubic decimeters

The fanga is also considered to equal 50.24 dry quarts.

1 BOTA

459.99200 (a) quarts (liquid)
26.00000 (e) almudes
435.3025214 (a) liters
435.3147632 (a) cubic decimeters

1 MOIO

754.56000 (a) quarts (dry)
60.00000 (e) alqueires
23.58000 (a) bushels
830.9373178 (a) liters
830.9605582 (a) cubic decimeters

The moio is also considered to equal 755.2 dry quarts.

1 TONELADA

919.98400 (a) quarts (liquid)
2.00000 (e) botas
870.6050428 (a) liters
870.6295264 (a) cubic decimeters

The tonelada is also considered to equal 919.92 liquid quarts.

Capacity and volume units of Russia

Russia, as is true of most countries, is currently using the metric system, but the following units persist in many measuring instances.

---

1 CHARKA 0.032500 (a) quart (liquid)
0.0307556 (a) liter
0.0307565 (a) cubic decimeter

The charka is also considered to equal 2.0 chkaliks.

1 CHKALIK 2.00000 (e) charkas
0.065000 (a) quart (liquid)
0.0615112 (a) liter
0.0615129 (a) cubic decimeter

The chkalik is also considered to equal 0.5 charka.

1 TCHAST 6.67000 (a) cubic inches
0.099252 (a) quart (dry)
0.1092980 (a) liter
0.1093010 (a) cubic decimeter

1 BOUTYLKA 12.50000 (e) chkaliks
0.812500 (a) quart (liquid)
0.7688902 (a) liter
0.7689117 (a) cubic decimeter

The boutylka is used in the measure of wine.

1 TOOP 2.00000 (e) chkaliks
1.30000 (a) quarts (liquid)
1.2302240 (a) liters
1.2302580 (a) cubic decimeters

The toop is used primarily in Estonia.

1 POLUGARNETZ 15.00000 (e) tchasts
1.48877 (a) quarts (dry)
1.6394699 (a) liters
1.6395151 (a) cubic decimeters

1 GARNETZ 2.97755 (a) quarts (dry)
2.00000 (e) polugarnetz

3.2789398 (a) liters
3.2790301 (a) cubic decimeters

The garnetz is also considered to equal 2.96 **dry** quarts.

1 VEDRO

16.00000 (e) boutylkas
13.00000 (a) quarts (liquid)
12.3022432 (a) liters
12.3025877 (a) cubic decimeters

1 CHETVERIK

23.82038 (a) quarts (dry)
8.00000 (e) garnetz
26.2315186 (a) liters
26.2322410 (a) cubic decimeters

The chetverik is also considered to equal 23.84 **dry** quarts.

1 PAJAK,
POLUOSMINA

47.64077 (a) quarts (dry)
2.00000 (e) chetveriks
1.48877 (a) bushels
52.4630371 (a) liters
52.4644820 (a) cubic decimeters

The unit is also considered to equal 47.68 dry quarts.

1 LOF, LOOF

592.00000 (e) tchasts
58.75695 (a) quarts (dry)
64.7044101 (a) liters
64.7061945 (a) cubic decimeters

The unit is also considered to equal 58.88 dry quarts.

1 OSMIN,
OSMINA

95.28154 (a) quarts (dry)
2.97755 (a) bushels
2.00000 (e) pajaks
104.9260742 (a) liters
104.9289641 (a) cubic decimeters

The unit is also considered to equal 95.36 dry quarts.

1 KOREC   111.95581 (a) quarts (dry)
                    23.50000 (e) pajaks
                    3.49862 (a) bushels
                    123.2881372 (a) liters
                    123.2915328 (a) cubic decimeters

1 CHETVERT   190.56307 (a) quarts (dry)
                    5.95510 (a) bushels
                    2.00000 (e) osmins
                    209.8521484 (a) liters
                    209.8579282 (a) cubic decimeters

The chetvert is also considered to equal 190.624 dry quarts.

1 BOTCHKA,   520.00000 (a) quarts (liquid)
FASS   40.00000 (e) vedro
                    492.0897280 (a) liters
                    492.1035080 (a) cubic decimeters

Capacity and volume units of Spain

Spain, as is true of most countries, is currently using the metric system, but the following units persist in many measuring instances.

1 COPA,   0.133000 (a) quart (liquid)
CUARTERON   0.1258614 (a) liter
                    0.1258649 (a) cubic decimeter

The copa is used in the measure of wine; the cuarteron in the measure of oil.

1 OCTAVILLO   0.250000 (e) cuartillo (dry)
                    0.2894375 (a) liter

1 CUARTILLO   8.00000 (e) copas
(LIQUID)   1.06400 (a) quarts (liquid)

1.0068913 (a) liters
1.0069195 (a) cubic decimeters

1 CUARTILLO
(DRY)

1.05133 (a) quarts (dry)
1.1577500 (a) liters
1.1577824 (a) cubic decimeters

1 AZUMBRE

2.12800 (a) quarts (liquid)
2.00000 (e) cuartillos (liquid)
2.0137826 (a) liters
2.0138390 (a) cubic decimeters

The azumbre is also considered to equal 4.0 liquid cuartillos. The unit is also used in Colombia and Panama.

1 MEDIO

2.10267 (a) quarts (dry)
2.00000 (e) cuartillos (dry)
2.3155000 (a) liters
2.3155647 (a) cubic decimeters

1 CUARTILLA
(LIQUID)

4.25600 (a) quarts (liquid)
2.00000 (e) azumbres
4.0275652 (a) liters
4.0276779 (a) cubic decimeters

1 CELEMIN

4.20533 (a) quarts (dry)
2.00000 (e) medios
4.6309999 (a) liters
4.6311294 (a) cubic decimeters

1 CUARTILLA
(DRY)

12.61600 (a) quarts (dry)
3.00000 (e) celemines
1.57700 (a) pecks
13.8929997 (a) liters
13.8933883 (a) cubic decimeters

The dry cuartilla is also considered to equal 12.56 dry quarts.

1 ARROBA   17.02400 (a) quarts (liquid)
4.25600 (a) gallons
4.00000 (e) cuartillas (liquid)
16.1102606 (a) liters
16.1107118 (a) cubic decimeters

The arroba is also considered to equal 4.263 gallons
and 16.1367578 liters. The unit equals and is known
as a centaro in Venezuela. The unit is also used in
Cuba. The term arroba also occurs under Weight and
Mass Units of Spain in Part Four and under Spirits
Measure in Part Three.

1 FANEGA   50.46400 (a) quarts (dry)
(DRY)   4.00000 (e) cuartillas (dry)
1.57700 (a) bushels
55.5719989 (a) liters
55.5735532 (a) cubic decimeters

The dry fanega is also used in Guatemala.

1 FANEGA   64.00000 (a) quarts (liquid)
(LIQUID)   60.5648896 (a) liters
60.5665856 (a) cubic decimeters

1 MOYO   272.38400 (a) quarts (liquid)
16.00000 (e) arrobas
257.7641701 (a) liters
257.7713883 (a) cubic decimeters

The moyo is also considered to equal 272.76 liquid
quarts.

1 CAFIZ, CAHIZ 605.56800 (a) quarts (dry)
18.92400 (a) bushels
12.00000 (e) fanegas (dry)
666.8639871 (a) liters
666.8826386 (a) cubic decimeters

The unit is also considered to equal 604.8 dry quarts.

1 CARGA   2.2200000 (a) hectoliters
2.2200622 (a) decisteres

The carga is used in the measure of grain, mostly in Castile.

Capacity and volume units of Sweden

Sweden, as is true of most countries, is currently using the metric system, but the following units persist in many measuring instances.

1 JUMFRU   0.086375 (a) quart (liquid)
0.0817389 (a) liter
0.0817412 (a) cubic decimeter

1 STOP   16.00000 (e) jumfrus
1.38200 (a) quarts (liquid)
1.3078231 (a) liters
1.3078597 (a) cubic decimeters

1 KANNOR   2.76400 (a) quarts (liquid)
2.00000 (e) stops
2.6156462 (a) liters
2.6157194 (a) cubic decimeters

1 KAPPE   4.83700 (a) quarts (liquid)
1.75000 (e) kannor
4.5773808 (a) liters
4.5775090 (a) cubic decimeters

1 FJARDING   16.62651 (a) quarts (dry)
7.00000 (e) kannor
18.3095232 (a) liters
18.3100359 (a) cubic decimeters

The fjarding is also considered to equal 16.64 dry quarts.

1 SPANN   66.50606 (a) quarts (dry)
4.00000 (e) fjardings
2.07831 (a) bushels
73.2380928 (a) liters
73.2401436 (a) cubic decimeters

The spann is also considered to equal 66.56 dry quarts.

1 TUNNA   133.01211 (a) quarts (dry)
4.15663 (a) bushels
2.00000 (e) spanns
146.4761855 (a) liters
146.4802873 (a) cubic decimeters

The tunna is also considered to equal 133.12 dry quarts.

1 AM   165.84000 (a) quarts (liquid)
60.00000 (e) kannor
156.9387702 (a) liters
156.9431649 (a) cubic decimeters

The am is also considered to equal 165.92 liquid quarts.

1 KOLTUNNA   149.63863 (a) quarts (dry)
9.00000 (e) fjardings
4.67621 (a) bushels
164.7857087 (a) liters
164.7903232 (a) cubic decimeters

The koltunna is also considered to equal 149.76 dry quarts.

1 OXHUVUD   248.76000 (a) quarts (liquid)
1.50000 (e) ams
235.4081553 (a) liters
235.4147474 (a) cubic decimeters

1 FODER   855.68000 (a) quarts (dry)
26.74000 (a) bushels

942.2927853 (a) liters
942.8026022 (a) cubic decimeters

1 KOLLAST
1,795.66351 (a) quarts (dry)
56.11449 (a) bushels
27.00000 (e) spanns
1,977.4285042 (a) liters
1,977.4838782 (a) cubic decimeters

The kollast is also considered to equal 1,795.2 dry quarts.

Capacity and volume units of Switzerland

Switzerland, as is true of most countries, is currently using the metric system, but the following units persist in many measuring instances.

1 POT, MAASS, IMMI
1.59000 (a) quarts (liquid)
1.5046590 (a) liters
1.5047011 (a) cubic decimeters

The unit is known as an immi if equaling 1.50 liters; the maass is also considered to equal 1.58 liquid quarts.

1 SETIER
9.54000 (a) gallons
6.00000 (a) pots
9.0279539 (a) liters
9.0282067 (a) cubic decimeters

The setier is also considered to equal 9.91 gallons.

1 VIERTEL, MUID
5.29486 (a) cubic feet
0.426000 (a) bushel
14.6589525 (a) liters
14.6598630 (a) cubic decimeters

The muid is also considered to equal 0.425 bushel, but the muid of South Africa equals 4.0 shepels and 3.1 bushels.

1 HOLZKLAFTER 105.89725 (a) cubic feet
20.00000 (a) viertels
293.1790501 (a) liters
293.1972596 (a) cubic decimeters

The holzklafter is also considered to equal 103 cubic feet.

1 MOULE 141.21000 (a) cubic feet
3,998.6907765 (a) liters
3,998.8027398 (a) cubic decimeters

Additional ancient and foreign capacity and volume units

Most of the countries or territories represented below now use the metric system, but the units mentioned still persist in many measuring instances.

FINLAND **1 kannu:** 2.76 quarts (liquid
1 ottinger: 4.15 gallons
1 tunna (dry): 149.728 quarts (dry), 4.679 bushels
1 tunna (liquid): 63.0 (e) kannu, 43.47 gallons; also considered to equal 33.19 gallons

HUNGARY **1 itcze:** 0.897 quart (liquid)
1 metze: 48.32 quarts (dry), 1.51 bushels; in Bavaria, 32 dreissigers, 8 massels, and 1.05 bushels
1 ako, eimer: 64.0 (e) itczes, 57.408 quarts (liquid), 54.32671 liters

PHILIPPINES **1 apatan:** 0.094 liter, 0.17 pint (dry)
1 chupa: 4.0 (e) apatans, 0.68 pint (dry)
1 ganta: 8.0 (e) chupas, 2.72 quarts (dry)
1 caban, cavan: 25.0 (e) gantas, 2.144 bushels; also considered to equal 2.13 bushels

VENEZUELA **1 centaro, arroba:** 4.263 gallons
1 fanega: 3.334 bushels; in Uruguay, 3.888 bushels; in

Chile, 2.753 bushels; in Ecuador and El Salvador, 1.575 bushels

MISCELLANEOUS
FOREIGN UNITS

1 achane (*ancient Persia*): 66.0 bu, 2,330 liters

1 adhaka (*India*): ¼ drona, 4.4 quarts (liquid)

1 adoulie (*Bombay*): 1/16 parah, 7.0 liters

1 akov (*Yugoslavia*): 40 okas, 15 gallons

1 amunam (*Ceylon*): 8 parahs, 203.4 liters

1 anoman (*Ceylon*): 5.77 U.S. bushels

1 bocoy (*Cuba*): 175 gallons

1 bodge (*obsolete, England*): 0.5 peck

1 boisseau (*Belgium*): 1.7 pecks

1 botella (*liquid*) (*El Salvador*): 0.77 quart (liquid)

1 bu (*ancient Persia*): 35.3 liters

1 caba (*Somaliland*): 0.479 quart (liquid)

1 caffiso (*Malta*): 5.4 gallons

1 cafiz, cahiz (*Nicaragua*): 17.24 bushels

1 cajuela (*Costa Rica*): 3.74 gallons

1 caneca (*Cuba*): 4.784 gallons

1 capicha (*Iran*): 2.63 liters

1 chenica (*Iran*): 1.359 liters

1 colluthun, colothun (*Persia*): ⅛ artaba, 0.233 bushel

1 coyang (*East Indies*): 3,561.0 liters

1 cuarta (*Paraguay*): 0.757 liter

1 dimerlie (*Rumania*): 22.34 quarts (dry), 24.6 liters

1 drona (*India*): 17.4 quarts (liquid)

1 eimer (*East Germany, West Germany*): 29 to 307 liters

1 fortin (*Turkey*): 4.0 (e) kileh, 4 bushels

1 fuder (*Luxembourg*): 264.18 gallons

1 fuder (*West Germany*): 220.0 gallons

1 garava (*Syria*): 41.15 bushels

1 hu (*Mainland China*): 51.77 liters

1 kantang (*Cambodia*): 1.65 gallons

1 kapetis (*ancient Persia*): 74.5 cubic inches, 1.2 liters

1 koilon (*Greece*): 33.17 liters

1 krina (*grain*) (*Bulgaria*): 18.16 quarts (dry)

1 kulmet (*Latvia*): 10.93 liters

Miscellaneous   **1 kwien** *(Thailand)*: 440 gallons
foreign units   **1 leaguer** *(South Africa)*: 127 gallons
(continued)   **1 lestrad, listred** *(Wales)*: 2.84 bushels
1 parah, para, parrah *(Ceylon)*: 23.1 quarts (dry)
1 parah, para, parrah *(Bombay)*: 100 quarts (dry)
1 parah, para, parrah *(Madras)*: 55.68 quarts (dry)
1 saa *(ancient Arabia)*: 1/12 cafiz, 2.91 quarts (liquid)
1 sheng *(Mainland China)*: 1.035 liters
1 skieppe *(Norway)*: 15.78 quarts (dry)
1 stero *(Italy)*: 35.31 cubic feet
1 takar *(East Indies)*: 6.81 gallons
1 tarri *(Algeria)*: 0.56 bushel
1 teman *(Arabia)*: 2.41 bushels
1 thangsat *(Thailand)*: 21.3 liters
1 toembak *(East Indies)*: 8.742 cubic yards
1 trug *(wheat)* *(old England)*: 0.67 bushel

PART FOUR

Units of weight and mass

Weight and mass units are presented in Part Four:

Troy weight

Avoirdupois weight

Apothecaries' weight

Intermeasure comparison of weight and mass units

Units of ancient Greece, India, and Rome

Units of Denmark, Egypt, Greece, Japan, the Netherlands, Poland, Portugal, Russia, Spain, Sweden, Yugoslavia, and other foreign units, both current and ancient

All the units described in Part Four, except where specified and through common usage, are considered units of weight, even though some may have been designed as units of mass. The term gravimetry is used to indicate the measurement of weight or density.

Weight is defined as the measured heaviness of a specific object, the gravitational force exerted by the earth or other celestial body on an object equal to the product of the object's mass and the local value of gravitational acceleration.

Mass is defined as a unit's physical volume of bulk, the measure of a body's resistance to acceleration. The mass of a body is different from, but proportional to, its weight; is independent of the body's position but dependent on its motion with respect to other bodies; and is expressed by means of the mass-energy relation of the special theory of relativity.

Troy weight

In 1266, Henry III decreed that "an English penny, called a sterling, round and without any clippings, shall weigh 32 wheatcorns in the middle of the ear. Twenty pence do make an ounce, and 12 ounces a pound." The English kings used troy (named for the French town Troyes) weight to count the royal treasures, and in 1527 it became legal standard for minting coins.

Any quantity of gold is supposed to be divided into 24 parts, or *carats*, and if pure, would be termed 24-carat fine. The new standard, used for watchcases, etc., is 18-carat fine. In the weighing of diamonds, the term carat applies to a weight of 3.2 grains troy. This carat is divided into four parts or grains; thus 4 grains troy equals 5 grains diamond weight.

The *beqa*, an ancient Egyptian unit equaling from less than 0.5 ounce to 7.5 ounces, is the earliest known standard of weight, dating from about 3800 B.C., and is still generally used in the weighing of precious metals, gems, and stones in troy weight. The ancient cylindrical standards varied from 188.7 to 211.2 grams.

1 DYNE

0.001020 (a) grain
0.0660949 (a) milligram

The term dyne is also used in Force Units, Part Seven.

1 POINT

2.0000000 (e) milligrams

The point is the jeweler's unit of mass. The term point is also used in Linear Measure and Printer's Measure, Part One, and in Angular Measure, Part Two.

1 MITE

49.01959 (r) dynes
0.050000 (e) grain
3.2399455 (r) milligrams

The mite is an ancient British unit and was the name of an ancient coin.

1 QUARTER,   16.19973 (r) mites
CARAT GRAIN,   0.771618 (r) grain
PEARL GRAIN   0.250000 (e) carat
                    50.0000000 (e) milligrams

1 GRAIN   980.39118 (r) dynes
                    20.00000 (e) mites
                    1.29598 (r) quarters
                    0.308647 (r) carat
                    64.7989100 (e) milligrams

The term grain is also used in Avoirdupois Weight and Apothecaries' Weight, Part Four. In old England, the grain was also known as a barleycorn, and the term barleycorn occurs also under Linear Measure in Part One.

1 CARAT, KARAT,   64.79891 (e) mites
INTERNATIONAL   4.00000 (e) quarters
CARAT, METRIC   3.23995 (r) grains
CARAT   0.134998 (r) pennyweight
                    200.0000000 (e) milligrams

The carat was derived from the weight of a seed of the carob tree. The original U.S. carat equaled 205.3 milligrams, but in 1913 the carat defined above was adopted.

1 PENNYWEIGHT   31.10348 (r) quarters
                    24.00000 (e) grains
                    7.77587 (r) carats
                    1.5551738 (r) grams

1 ASSAY TON   583.33333 (r) quarters
                    450.10983 (r) grains
                    145.83333 (r) carats
                    18.75458 (r) pennyweights

| *1 Assay ton* | 0.937730 (r) ounce |
| *(continued)* | 29.1666667 (r) grams |

The assay ton is used for testing ore, and its value is determined by the following relation:

$$\frac{\text{assay ton}}{\text{milligrams}} = \frac{\text{short ton (avoirdupois)}}{\text{ounces (troy)}}$$

1 OUNCE

9,600.00000 (e) mites
622.06954 (r) quarters
480.00000 (e) grains
155.51738 (r) carats
20.00000 (e) pennyweights
1.06641 (r) assay tons
0.500000 (e) mancus
0.083333 (r) pound
31.1034768 (e) grams

The ounce has also been considered to equal 150.0 (e) carats. The term ounce also occurs under Linear Measure in Part One, Spirits Measure in Part Three, and Avoirdupois Weight and Apothecaries' Weight in Part Four.

1 MANCUS

960.00000 (e) grains
311.03478 (r) carats
2.00000 (e) ounces
62.2069536 (e) grams

The mancus is an ancient and obsolete British unit.

1 POUND

5,760.00000 (e) grains
1,866.20861 (r) carats
240.00000 (e) pennyweights
12.79685 (r) assay tons
12.00000 (e) ounces
6.00000 (e) mancus
373.2417216 (e) grams

The term pound also occurs under Avoirdupois Weight and Apothecaries' Weight in Part Four and under Force Units in Part Seven.

1 MERCHANT'S POUND
7,680.00000 (e) grains
1.33333 (r) pounds
497.6556288 (e) grams

1 MAST
2.50000 (e) pounds
933.1043040 (e) grams

The mast is an obsolete English unit.

Avoirdupois weight

Edward I (1239–1307) of England made many administrative changes and improvements in a chaotic reign that began in 1272. One of these changes was to recognize and legalize, in 1303, such units as the ounce, pound, wey, stone, and hundredweight, and probably many others as well.

These weight units were eventually, in 1335, grouped together under the name *avoirdupois*, a French word meaning "goods of weight," meant primarily to be used in commerce.

1 ATOMIC MASS UNIT
1.66043×10^{-27} (r) kilogram

1 GRAIN
0.036571 (r) dram
64.7989100 (e) milligrams

The grain is known as a grein in the Netherlands. The term grain is also used in Troy Weight and Apothecaries' Weight, Part Four.

1 DRAM
27.34375 (e) grains
0.062500 (e) ounce
1.7718452 (r) grams

The dram was originally a Roman weight called a dracham, which meant 1/12 part and equaled 1/12 pound, but the term may have originated from the ancient Greek word *drachm*, meaning "a handful."

1 Dram *(continued)*	The term dram also occurs under Apothecaries' Weight in Part Four.

1 OUNCE

437.50000 (e) grains

16.00000 (e) drams

0.062500 (e) pound

28.3495231 (r) grams

The ancient Romans divided the pound into 12 ounces, each equal to 437.0 grains. The English later increased the pound to 16 ounces, or a total of 6,992 grains, which was rounded off to 7,000 grains, making each new ounce now 437.5 grains.

The ounce is also considered to equal 28.349527 grams. The term ounce also occurs under Linear Measure in Part One, Spirits Measure in Part Three, and Troy Weight and Apothecaries' Weight in Part Four.

1 POUND

7,000.00000 (e) grains

256.00000 (e) drams

16.00000 (e) ounces

0.050000 (e) score

453.5923700 (e) grams

The avoirdupois pound was established by international agreement in July 1959, as defined above, to replace the U.S. avoirdupois pound of 453.5924277 grams and the U.K. pound of 453.592338 grams. The term pound also occurs under Troy Weight and Apothecaries' Weight in Part Four and under Force Units in Part Seven.

1 CATTY

21.33333 (r) ounces

1.33333 (r) pounds

6.0477471 (r) hectograms

The catty is used to measure tea and rice in eastern Asia, and may vary with locality: 1.33 pounds in China and Malaysia; 1.36 pounds in Java; 1.32 or 2.67 pounds in Siam; 2.12 pounds in Sumatra; and 2.67 pounds in Thailand. The pound of Japan, known as a kin, equals 1.32 pounds.

1 QUARTERN   64.00000 (e) ounces
4.00000 (e) pounds
0.200000 (e) score
1.8143695 (r) kilograms

The quartern is used to measure bread and is also considered to equal ¼ stone and ¼ avoirdupois pound. The term quartern also occurs under Liquid Measure and Dry Measure in Part Three.

1 BLOCK   80.00000 (e) ounces
5.00000 (e) pounds
0.500000 (e) gallon
2.2679619 (r) kilograms

The block is used in the measure of cotton. The term block also occurs under Linear Measure in Part One and under Square Measure in Part Two.

1 HEAD   108.00000 (e) ounces
6.75000 (e) pounds
3.0617485 (r) kilograms

1 CLOVE, BRICK   112.00000 (e) ounces
7.00000 (e) pounds
0.500000 (e) stone
3.1751466 (r) kilograms

The clove, a British unit, is also considered to vary from 8 to 10 avoirdupois pounds in the measure of cheese and wool.

1 GALLON   160.00000 (e) ounces
10.00000 (e) pounds
0.500000 (e) score
4.5359237 (e) kilograms

The term gallon also occurs under Liquid Measure, Apothecaries' Fluid Measure, Cooking Measure, and Spirits Measure in Part Three.

1 STONE, PECK   224.00000 (e) ounces
14.00000 (e) pounds
2.00000 (e) cloves
0.500000 (e) British quarter
6.3502932 (r) kilograms

The stone, a British unit, weighs 16.0 (e) pounds if used to measure wool. The peck was originally used to measure flour. The term peck also occurs under Dry Measure and Cooking Measure in Part Three.

1 SCORE   320.00000 (e) ounces
20.00000 (e) pounds
2.00000 (e) gallons
9.0718474 (e) kilograms

The score is a British unit used in the measure of wool. The term score also occurs under Units of Count in Part Six.

1 QUARTER   400.00000 (e) ounces
25.00000 (e) pounds
5.00000 (e) blocks
2.50000 (e) gallons
0.892857 (r) British quarter
0.500000 (e) frail
11.3398093 (r) kilograms

The quarter is used to measure grain, but may also equal ¼ pound, ¼ ton, etc. The term quarter also occurs under Cloth Measure in Part One, Dry Measure in Part Three, Troy Weight in Part Four, and Time Measure in Part Six.

1 BRITISH   448.00000 (e) ounces
QUARTER, TOD   28.00000 (e) pounds
7.00000 (e) quarterns
4.00000 (e) cloves
2.00000 (e) stones
1.12000 (e) quarters

0.500000 (e) firkin

12.7005864 (r) kilograms

The British quarter is used in the measure of wool.

1 SLUG   32.17000 (a) pounds

14.5920665 (a) kilograms

The term slug also occurs under Force Units in Part Seven.

1 TRUSS   36.00000 (e) pounds
(STRAW)   9.00000 (e) quarterns

0.642857 (r) truss (old hay)

0.600000 (e) truss (new hay)

0.027778 (r) load (straw)

16.3293253 (r) kilograms

The truss is a British unit. A similar weight is the pood of Russia, which equals 36.113 pounds.

1 FRAIL   50.00000 (e) pounds

12.50000 (e) quarterns

10.00000 (e) blocks

5.00000 (e) gallons

2.50000 (e) scores

2.00000 (e) quarters

0.500000 (e) hundredweight

22.6796185 (e) kilograms

The frail, a unit of Spain, is used to measure raisins. The unit received its name from a rush basket used in Mediterranean countries for raisins and figs. A similar weight is the tambor of Bolivia, which also equals 50 pounds.

1 FIRKIN, TRUSS   56.00000 (e) pounds
(OLD HAY)   2.00000 (e) British quarters

1.55556 (r) trusses (straw)

0.933333 (r) truss (new hay)

0.500000 (e) long hundredweight

1 Firkin, truss *(old hay)* *(continued)*	0.027778 (r) load (old hay) 25.4011727 (r) kilograms

The firkin is a British unit used to measure lard or butter. The term firkin also occurs under Spirits Measure in Part Three.

1 TALENT

The talent was an ancient unit, often equaling 60 mina, but varying with time and place. The later Attic talent is estimated at 58 avoirdupois pounds, and the Hebrew talent about double this.

The mina was an ancient unit of both weight and value, probably of Babylonian origins.

1 TRUSS (NEW HAY)

60.00000 (e) pounds
1.66667 (r) trusses (straw)
1.07143 (r) trusses (old hay)
27.2155422 (e) kilograms

1 FRENCH FOOTWEIGHT

62.42450 (r) pounds
0.985842 (r) footweight
28.3152769 (r) kilograms

1 BUSHEL

63.00000 (e) pounds
28.5763193 (r) kilograms

The bushel is a British unit. The term bushel also occurs under Dry Measure and Cooking Measure in Part Three.

1 FOOTWEIGHT, TALENT

63.32100 (r) pounds
1.01436 (r) French footweights
28.7219225 (r) kilograms

The footweight is a British unit.

1 FLASK

75.00000 (e) pounds
0.986842 (r) British flask
34.0194278 (r) kilograms

The flask is used to measure mercury.

1 BRITISH FLASK   76.00000 (e) pounds
1.01333 (r) flasks
34.4730201 (r) kilograms

1 TUB   84.00000 (e) pounds
0.500000 (e) sack
38.1017591 (r) kilograms

The tub defined above is a British unit. In the United States a tub varies from 0.5 to 3.0 bushels of oysters.

1 BOX   90.00000 (e) pounds
40.8233133 (e) kilograms

The box is a British unit.

1 HUNDRED-   100.00000 (e) pounds
WEIGHT   25.00000 (e) quarterns
20.00000 (e) blocks
10.00000 (e) gallons
5.00000 (e) scores
4.00000 (e) quarters
2.00000 (e) frails
0.892857 (r) long hundredweight
4.5359237 (e) myriagrams

The hundredweight is also known as a cental, but if used to measure nails it is known as a keg. The term keg also occurs under Spirits Measure in Part Three.

1 QUINTAL   The quintal varies with locality: 100 pounds in El Salvador, 101.3 pounds in Argentina, 101.43 pounds in Peru and Chile, 101.47 pounds in Mexico, 129.54 pounds in Brazil, and 220.5 pounds in Angola. The term quintal is also used in Metric Weight and Mass Units, Part Five.
The quinta, a similar unit, equals 101.467 pounds in Mexico, 110.2 pounds in Switzerland, and 110.231 pounds in Colombia.

1 CENTNER   The centner varies with locality: 93.7 pounds in Sweden, 110.231 pounds in Denmark and Norway, 113.44 pounds in Germany, and 117.5 pounds in

1 Centner *(continued)*	Brunswick. The centner is also considered to equal 50.0 (e) kilograms in Austria, Denmark, and Switzerland, and with this value is also known as a metric hundredweight.

1 MAIL

105.00000 (e) pounds

4.7627199 (r) myriagrams

The mail is a Scottish unit.

1 LONG HUNDREDWEIGHT

112.00000 (e) pounds

8.00000 (e) cloves

4.00000 (e) British quarters

2.00000 (e) firkins

1.12000 (e) hundredweights

0.666667 (r) sack

5.0802345 (r) myriagrams

It was Queen Elizabeth I of England who gave 12 pounds more to the hundredweight.

1 FAGOT, SEAM

120.00000 (e) pounds

24.00000 (e) blocks

12.00000 (e) gallons

1.33333 (r) boxes

0.500000 (e) pack (wool and flax)

5.4431084 (r) myriagrams

The fagot is a British unit used to measure iron and steel; the seam, also a British unit, is used to measure glass. The term seam also occurs under Dry Measure in Part Three.

1 BOLL

140.00000 (e) pounds

5.00000 (e) British quarters

0.500000 (e) pack (meal)

6.3502932 (r) myriagrams

The boll is a Scottish unit used to measure oatmeal. The term boll is also used in Dry Measure, Part Three.

1 SACK, POCKET   168.00000 (e) pounds

2.00000 (e) tubs

1.50000 (e) long hundredweights

7.6203518 (r) myriagrams

The sack is a British unit used in the measure of coal, potatoes, and wool. If used to measure cotton, a sack weighs 140 pounds, equal to the boll; a sack of salt weighs 315 pounds. A sack of flour weighs 100 pounds if used in the United States, but 140 pounds if exported from the United States. The pocket is an English unit used in the measure of hops.

An obsolete British unit, the quarter sack, weighs 91.0 pounds or ¼ bag.

1 STAND   The stand is an English unit varying from 2.5 to 3.0 hundredweight of pitch. In South Africa it is a relatively small piece of land (a surface unit) measured off by an official surveyor for sale or building.

1 SHIP POUND   The ship pound, used in northern Europe, varies between 300 and 400 pounds and commonly contains 20 lispunds, itself a varying weight unit.

1 WEY   182.00000 (e) pounds

0.500000 (e) bag

8.2553811 (r) myriagrams

The wey is a British unit, and if used to measure wool is known as a pocket. The term wey also occurs under Dry Measure in Part Three.

1 BARREL   196.00000 (e) pounds

8.8904105 (r) myriagrams

The barrel as defined above is used to measure flour, but a barrel weighs 200 pounds when used to measure fish, beef, and pork, while a barrel of lime varies from 180 to 280 pounds. A barrel of cement weighs 376 pounds or 4 bags as used in Dry Measure, Part Three. A barrel containing fruit or vegetables must contain at least 7,056 cubic inches of produce.

The term barrel also occurs under Liquid Measure, Dry Measure, and Spirits Measure in Part Three.

---

1 PACK (WOOL
AND FLAX)

240.00000 (e) pounds
2.00000 (e) fagots
0.857143 (r) pack (meal)
10.8862169 (r) myriagrams

1 PACK (MEAL)

280.00000 (e) pounds
1.16667 (r) packs (wool and flax)
12.7005864 (r) myriagrams

1 PIG

301.00000 (e) pounds
13.6531303 (r) myriagrams

The pig is a British unit used to measure ballast.

1 TIERCE

320.00000 (e) pounds
14.5149558 (r) myriagrams

The tierce is a British unit. The term tierce occurs
also under Spirits Measure in Part Three.

1 BAG

364.00000 (e) pounds
2.00000 (e) weys
16.5107623 (r) myriagrams

The bag is used to measure wool; in Britain, however,
a bag of wool, called a pocket, weighs 182 pounds,
equal to the wey. A bag of cement weighs 94 pounds
net. For coffee, a bag in Brazil weighs 132 pounds net;
in Colombia and Costa Rica, 140 pounds gross; in
Mexico, Nicaragua, and Venezuela, 145 pounds gross;
and in Guatemala, 160 pounds gross. The term bag
also occurs under Dry Measure in Part Three.

1 BALE

500.00000 (e) pounds
22.6796185 (e) myriagrams

The bale is used to measure cotton, but varies with
locality: in Brazil and Peru, 250 pounds; in India,
400 pounds; in Egypt, 740 pounds. In the United
States, the bale is 500 pounds as established by the
Cotton Futures Act Number 39, but 477 pounds may
also be in common use. The bale was so named because
cotton was baled or tied into big bundles for storing

---

or shipping. The term bale also occurs under Paper Measure in Part Six.

1 CASK

672.00000 (e) pounds

30.4814073 (r) myriagrams

The term cask also occurs under Spirits Measure in Part Three.

1 KIP

1,000.00000 (e) pounds

4.5359237 (e) quintals

The term kip is derived from the first two letters of kilogram and the first letter of pound.

1 HOGSHEAD

1,200.00000 (e) pounds

5.4431084 (e) quintals

The term hogshead also occurs under Liquid Measure and Spirits Measure in Part Three.

1 CISTERN

1,600.00000 (e) pounds

5.00000 (e) tierces

7.2574779 (r) quintals

1 LOAD

1,800.00000 (e) pounds

8.1646627 (r) quintals

The load, a British unit, varies with different commodities: a load of ore weighs 576 pounds; straw, 1,296 pounds and 36 trusses of straw; old hay, 2,016 pounds and 36 trusses of old hay; new hay, 2,160 pounds and 36 trusses of new hay. In Sierra Leone, a load equals 60 pounds of cocoa. The term load also occurs under Cubic Measure and Dry Measure in Part Three.

1 FODDER

1,950.00000 (e) pounds

8.8450512 (r) quintals

The fodder is a British unit.

1 SHORT TON,	2,000.00000 (e) pounds
NET TON	80.00000 (e) quarters
	20.00000 (e) hundredweights
	2.00000 (e) kips
	0.892857 (r) long ton
	0.500000 (e) last
	9.0718474 (e) quintals

The short ton is also considered to equal 0.9071849 metric ton. A short ton of soft coal occupies about 45 cubic feet of space, of coke about 65 cubic feet, and of hard coal about 35 to 35.25 cubic feet.

A kiloton is a unit of explosive force equal to 1,000 short tons of TNT; a megaton equals 1,000,000 short tons of TNT.

1 TONNE	2,204.62261 (r) pounds
	10.0000000 (e) quintals

The tonne is a French unit.

1 LONG TON,	2,240.00000 (e) pounds
GROSS TON,	80.00000 (e) British quarters
BRITISH TON	20.00000 (e) long hundredweights
	1.12000 (e) short tons
	10.1604691 (r) quintals

The long ton is a British unit; when used to measure wool it is called a sarpler.

1 CHALDRON	2,625.00000 (e) pounds
	11.9067997 (r) quintals

The term chauldron also occurs under Dry Measure in Part Three.

1 LAST	4,000.00000 (e) pounds
	160.00000 (e) quarters
	2.00000 (e) short tons
	18.1436948 (e) quintals

The last varies with different commodities: a last of flax or feathers equals 1,700 pounds; a last of gunpowder equals 24 barrels (Dry Measure, Part Three) weighing 100 pounds each.

A last in Germany equals 2.0 metric tons.

A last also varies with count and capacity: a last of wool equals 12 sacks; a last of leather hides is 12 dozen; a last of herring is 10,000, 13,200, or 20,000 individual fish.

The term last also occurs under Dry Measure in Part Three, and the unit is sometimes called a load, a term used elsewhere in Avoirdupois Weight.

1 ROOM

14,000.00000 (e) pounds

500.00000 (e) British quarters

6.3502932 (r) tonneaus

The room is a British unit used in the measure of coal.

1 KEEL

42,400.00000 (e) pounds

378.57143 (r) long hundredweights

18.92857 (r) long tons

19.2323165 (r) tonneaus

The keel is a British unit used in the measure of coal. The term keel also occurs under Cubic Measure in Part Three.

1 BARGE

47,040.00000 (e) pounds

420.00000 (e) long hundredweights

23.50000 (e) short tons

21.00000 (e) long tons

21.3369851 (r) tonneaus

The barge is used to measure coal, and is also considered to equal 4.0 hundredweights and 21.2 long tons of freight.

1 ROD WEIGHT

55,600.00000 (e) pounds

496.42857 (r) long hundredweights

27.80000 (e) short tons

24.82143 (r) long tons

25.2197358 (r) tonneaus

Apothecaries' weight

Until 1617, both grocers and apothecaries (pharmacists) in England could sell drugs and medicine. Then the apothecaries obtained a separate charter with the College of Physicians, and this decided the standards for measure. Today, the following units are still generally used in compounding medicines.

1 GRAIN

0.050000 (e) scruple
64.7989100 (e) milligrams

The term grain is also used in Troy Weight and Avoirdupois Weight in Part Four.

1 SCRUPLE, SCRUPPLE

20.00000 (e) grains
0.333333 (r) dram
0.041667 (r) ounce
1.2959782 (e) grams

The unit was originally a small pointed stone, used as a weight standard. In Russia a scruple equals 19.198 grains.

1 DRAM

60.00000 (e) grains
3.00000 (e) scruples
0.125000 (e) ounce
3.8879346 (e) grams

The term dram is also used in Avoirdupois Weight in Part Four.

1 OUNCE

480.00000 (e) grains
24.00000 (e) scruples
8.00000 (e) drams
31.1034768 (e) grams

The term ounce also occurs under Linear Measure in Part One, Spirits Measure in Part Three, and Troy Weight and Avoirdupois Weight in Part Four.

1 POUND   5,760.00000 (e) grains
                   288.00000 (e) scruples
                   96.00000 (e) drams
                   12.00000 (e) ounces
                   373.2417216 (e) grams

The term pound also occurs under Troy Weight and Avoirdupois Weight in Part Four and under Force Units in Part Seven.

Intermeasure comparison of weight and mass units

The following material shows the relation between various units of weight and mass. To identify the section in which a particular unit appears, the following code is used:

(T) Troy Weight

(V) Avoirdupois Weight

(P) Apothecaries' Weight

1 CARAT (T)   0.154324 (r) scruple (P)

1 SCRUPLE (P)   6.47989 (r) carats (T)
                       0.731429 (r) dram (V)

1 DRAM (V)   8.85923 (r) carats (T)
                       0.455729 (r) dram (P)

1 DRAM (P)   2.50000 (e) pennyweights (T)
                       2.19428 (r) drams (V)

1 OUNCE (V)   7.29167 (r) drams (P)
                       0.971983 (r) assay ton (T)

1 ASSAY TON (T)   16.46118 (r) drams (V)

---

1 OUNCE (T, P)   1.09714 (r) ounce (V)

1 POUND (T, P)   0.822857 (r) pound (V)

1 POUND (V)   1.21528 (r) pounds (T, P)

1 LONG HUN-   136.11111 (r) pounds (T, P)
DREDWEIGHT (V)

1 LONG TON (V)   2,722.22222 (r) pounds (T, P)

Weight and mass units of ancient Greece

The following units may prove useful to the archeologist and to students of the arts and history.

1 CHALCON   1.38000 (a) grains
                  89.4224958 (a) milligrams

1 OBOLOS   11.04000 (a) grains
                 8.00000 (e) chalca
                 715.3799664 (a) milligrams

The obolos, besides equaling 11.2 grains and 0.73
gram, also equals 11.08 grains

1 DIOBOLON   22.08000 (a) grains
                   2.00000 (e) oboloi
                   1.4307599 (a) grams

1 DRACHMA   66.24000 (a) grains
                  3.00000 (e) diobola
                  4.2922798 (a) grams

1 TETRA-   264.96000 (a) grains
DRACHMA   4.00000 (e) drachmai
                17.1691192 (a) grams

1 MINA   25.00000 (e) tetradrachmai
0.946300 (a) pound (avoirdupois)
429.2279798 (a) grams

Weight and mass units of ancient India

The following units may prove useful to the archeologist and to students of the arts and history.

1 YAVA   0.225000 (a) grain
14.5797548 (a) milligrams

1 RETTI   10.00000 (e) yavas
2.25000 (a) grains
145.7975475 (a) milligrams

1 KONA   108.00000 (a) grains
48.00000 (e) retti
0.246900 (a) ounce (avoirdupois)
6.9982823 (a) grams

1 KARSHA   216.00000 (a) grains
2.00000 (e) konas
0.493700 (a) ounce (avoirdupois)
13.9965646 (a) grams

1 PALA   728.00000 (a) grains
3.37037 (r) karshas
0.104000 (a) pound (avoirdupois)
47.1736065 (a) grams

Weight and mass units of ancient Rome

The following units may prove useful to the archeologist and to students of the arts and history.

1 SCRIPULUM,   17.46700 (a) grains
SCRUPULUS   1.1318426 (a) grams

1 SEXTULA,   69.88000 (a) grains
SOLIDUS   4.00000 (e) scrupula
                      4.5273702 (a) grams

                      The unit is also considered to have equaled 0.87 troy
                      ounce or 417.6 grains.

1 SICILICUM   104.82000 (a) grains
                      1.50000 (e) sextulae
                      6.7910554 (a) grams

1 DUELLA   139.76000 (a) grains
                      1.33333 (r) sicilica
                      9.0547405 (a) grams

1 UNCIA   419.28000 (a) grains
                      3.00000 (e) duellas
                      27.1642215 (a) grams

1 BES   3,354.24000 (a) grains
                      8.00000 (e) unciae
                      0.481300 (a) pound (avoirdupois)
                      217.3137717 (a) grams

1 DODRAN   3,773.52000 (a) grains
                      9.00000 (e) unciae
                      8.62500 (a) ounces (avoirdupois)
                      244.4779932 (a) grams

1 AS, LIBRA,   5,031.36000 (a) grains
PONDUS   1.50000 (e) besses
                      1.33333 (r) dodrans
                      325.9706576 (a) grams

Weight and mass units of Denmark

Denmark, as is true of most countries, is currently using the metric system, but the following units persist in many measuring instances.

1 ES
0.843111 (a) grain
54.6326868 (a) milligrams

1 ORT
9.15200 (e) es
7.71615 (a) grains
499.9983586 (a) milligrams

1 KVINT,
QUINTIN, QUINT
77.16154 (a) grains
10.00000 (e) ort
4.9999836 (a) grams

1 PUND
9,152.00000 (e) es
7,716.15385 (a) grains
100.00000 (e) kvint
1.10231 (a) pounds (avoirdupois)
499.9983586 (a) grams

The pund is also used in Iceland; and in Denmark it is also considered to equal 7,714 grains.

1 BISMERPUND
92,593.84615 (a) grains
13.22769 (a) pounds (avoirdupois)
12.00000 (e) pund
5.9999803 (a) kilograms

The bismerpund is also used in Norway; in Denmark it is also considered to equal 92,400 grains.

1 LISPUND,
LISPOUND
123,458.46154 (a) grains
17.63692 (a) pounds (avoirdupois)
16.00000 (e) pund
7.9999737 (a) kilograms

The unit is also considered to equal 123,480 grains.

1 VOG	277,781.53846 (a) grains
	39.68308 (a) pounds (avoirdupois)
	3.00000 (e) bismerpund
	17.9999409 (a) kilograms

The vog is also considered to equal 277,760 grains.

1 SKIPPUND	352.73846 (a) pounds (avoirdupois)
	320.00000 (e) pund
	159.9994748 (a) kilograms

1 SKIBSLAST	5,732.00000 (a) pounds (avoirdupois)
	5,200.00000 (e) pund
	2,599.9914648 (a) kilograms

Weight and mass units of Egypt

Egypt, as is true of most countries, is currently using the metric system, but the following units persist in many measuring instances.

| **1 KERAT** | 3.01000 (a) grains |
| | 195.0447191 (a) milligrams |

The kerat in Turkey, however, equals 1/16 dirhem and 3.09 grains.

| **1 KAT** | 144.00000 (a) grains |
| | 9.3310430 (a) grams |

The kat is an ancient and probably obsolete unit that is also considered to have equaled 146 grains.

1 OKIA, OKIEH	577.79167 (a) grains
	1.32067 (a) ounces (avoirdupois)
	37.4402702 (a) grams

The unit is also considered to equal 577.5 grains.

1 DEBEN   1,440.00000 (a) grains
10.00000 (e) kat
93.3104304 (a) grams

The deben is an ancient and probably obsolete unit that is also considered to have equaled 90.0 grams.

1 ARTAL   6,933.50000 (a) grains
12.00000 (e) okias
449.2832427 (a) grams

1 OKA, OKE   19,260.50000 (a) grains
2.75150 (a) pounds (avoirdupois)
1,248.0594061 (a) grams

The unit is also considered to equal 1,203.125 grams.

1 KHAR   3,600.00000 (e) kat
75.08571 (a) pounds (avoirdupois)
33.5917549 (a) kilograms

The khar is an ancient and probably obsolete unit that is also considered to have equaled 74.96 avoirdupois pounds.

1 KANTAR   693,350.00000 (a) grains
100.00000 (e) artals
99.05000 (a) pounds (avoirdupois)
44.9283243 (a) kilograms

The kantar varies with location: in Greece, a kantar equals 124.16 pounds; in Cyprus, 44 okas and 123.2 pounds; in Malta, 175 pounds; in Tunisia, 100 artals and 111.1 pounds; and in Turkey, 44 okas and 124.45 pounds.

1 HAMLAH   165.09000 (a) pounds (avoirdupois)
60.00000 (e) okas
74.8835644 (a) kilograms

1 HEML   550.30000 (a) pounds (avoirdupois)
                    200.00000 (e) okas
                    249.6118812 (a) kilograms

Weight and mass units of Greece

Greece, as is true of most countries, is currently
using the metric system, but the following units
persist in many measuring instances.

1 DRAMME   49.38000 (a) grains
                    3.1997702 (a) grams

1 OKA, OKE   400.00000 (e) drammes
                    2.82171 (a) pounds (avoirdupois)
                    1.2799081 (a) kilograms

                    The unit, as defined above, is also used in Yugo-
                    slavia. But in Turkey and Bulgaria the oka equals
                    2.83 pounds; in Egypt, 2.75 pounds; and in Cyprus,
                    2.8 pounds.

1 KANTAR,   124.16000 (a) pounds (avoirdupois)
STATER   56.3180287 (a) kilograms

                    The kantar in Cyprus equals 44 okas and 123.2
                    pounds; in Egypt, 100 artals and 99.05 pounds; in
                    Malta, 175 pounds; in Tunisia, 100 artals and 111.1
                    pounds; and in Turkey, 44 okas and 124.45 pounds.

1 TALANTON   330.69000 (a) pounds (avoirdupois)
                    149.9984608 (a) kilograms

1 TONOS   3,306.90000 (a) pounds (avoirdupois)
                    10.00000 (e) talantons
                    1,499.9846080 (a) kilograms

Weight and mass units of Japan

Japan, as is true of most countries, is currently using the metric system, but the following units persist in many measuring instances.

1 SHI
0.005787 (a) grain
0.3749848 (a) milligram

1 MO
10.00000 (e) shi
0.057869 (a) grain
3.7498481 (a) milligrams

1 RIN
10.00000 (e) mo
0.578690 (a) grain
37.4984812 (a) milligrams

1 FUN
10.00000 (e) rin
5.78690 (a) grains
374.9848123 (a) milligrams

1 MOMME
57.86900 (a) grains
10.00000 (e) fun
3.7498481 (a) grams

The momme is also considered to equal 57.97 grains.

1 NIYO
216.10000 (a) grains
14.0030445 (a) grams

1 HIYAKU-ME
5,786.90000 (a) grains
100.00000 (e) momme
374.9848123 (a) grams

1 KIN, KATI
9,259.04000 (a) grains
160.00000 (e) momme
1.32272 (a) pounds (avoirdupois)
599.9756960 (a) grams

The unit is also considered to equal 9,261 grains.

1 KWAN,	57,869.00000 (a) grains
KWAMME	1,000.00000 (e) momme
	8.26700 (a) pounds (avoirdupois)
	3.7498481 (a) kilograms

1 HIYAK-KIN,	925,904.00000 (a) grains
PICUL	132.27200 (a) pounds (avoirdupois)
	16.00000 (e) kwans
	59.9975696 (a) kilograms

The unit is also considered to equal 926,100 grains.

1 KOMMA-ICHI-	2,314,760.00000 (a) grains
DA	330.68000 (a) pounds (avoirdupois)
	40.00000 (e) kwans
	149.9939240 (a) kilograms

The komma-ichi-da is also considered to equal 2,314,-900 grains.

Weight and mass units of the Netherlands

The Netherlands, or Holland, as is true of most countries, is currently using the metric system, but the following units persist in many measuring instances.

| 1 KORREL | 1.54306 (a) grains |
| | 99.9887681 (a) milligrams |

1 WICHTJE	15.43063 (a) grains
	10.00000 (e) korrels
	99.9887681 (a) centigrams

The wichtje is also considered to equal 15.432 grains.

1 LOOD	154.30625 (a) grains
	10.00000 (e) wichtjes
	9.9988768 (a) grams

1 ONS   1,543.06250 (a) grains
10.00000 (e) loods
3.52700 (a) ounces (avoirdupois)
99.9887681 (a) grams

The ons is also considered to equal 771.4 grains.

1 POND   7,715.31250 (a) grains
5.00000 (e) onsen
1.10219 (a) pounds (avoirdupois)
499.9438403 (a) grams

The pond is also considered to equal 7,714 grains and 10 ons.

Weight and mass units of Poland

Poland, as is true of most countries, is currently using the metric system, but the following units persist in many measuring instances.

1 SKRUPUL 16.29688 (a) grains
1.0560458 (a) grams

1 LUT   195.56250 (a) grains
12.00000 (e) skrupuls
12.6725493 (a) grams

1 UNCYA   391.12500 (a) grains
2.00000 (e) luts
0.894000 (a) ounce (avoirdupois)
25.3450987 (a) grams

1 FUNT   6,258.00000 (a) grains
16.00000 (e) uncyas
0.894000 (a) pound (avoirdupois)
405.5115788 (a) grams

1 Funt *(continued)*	The funt is also considered to equal 405.504 grams, and in Russia a funt equals 0.903 avoirdupois pound.
1 KAMIAN	156,450.00000 (a) grains 25.00000 (e) funts 22.35000 (a) pounds (avoirdupois) 10.1377895 (a) kilograms

The kamian in Russia equals 22.57 avoirdupois pounds.

Weight and mass units of Portugal

Portugal, as is true of most countries, is currently using the metric system, but the following units persist in many measuring instances.

1 GRAO	0.768750 (a) grain 49.8141621 (a) milligrams
1 ESCROPULO	24.00000 (e) graos 18.45000 (a) grains 1.1955399 (a) grams
1 ONCA	442.80000 (a) grains 28.6929573 (a) grams

The onca, as defined above, is also used in Brazil and is also considered to equal 442.7 grains.

1 LIBRA	9,216.00000 (e) graos 7,084.80000 (a) grains 16.00000 (e) oncas 1.01211 (a) pounds (avoirdupois) 459.0873176 (a) grams

The libra, as defined above, is also used in Brazil. But in Argentina a libra equals 1.013 pounds; in Colombia, 1.102 pounds; in Mexico, 1.015 pounds;

and in Spain, Bolivia, Chile, Ecuador, Peru, Cuba, and Venezuela, 1.014 pounds.

1 ARROBA

32.38765 (a) pounds (avoirdupois)
32.00000 (e) libras
14.6907942 (a) kilograms

The arroba is also considered to equal 32.38 (a) avoirdupois pounds in Brazil, and 25.36 (a) avoirdupois pounds in Spain and Mexico.

Weight and mass units of Russia

Russia, as is true of most countries, is currently using the metric system, but the following units persist in many measuring instances.

1 DOLA

0.685113 (a) grain
44.6224305 (a) milligrams

1 ZOLOTNIK

96.00000 (e) doli
65.77083 (a) grains
4.2837533 (a) grams

The zolotnik is also considered to equal 65.83 grains.

1 LOT

197.31250 (a) grains
3.00000 (e) zolotniks
0.451000 (a) ounce (avoirdupois)
12.8512600 (a) grams

1 FUNT

6,314.00000 (a) grains
32.00000 (e) lots
0.902000 (a) pound (avoirdupois)
411.2403177 (a) grams

The funt is also considered to equal 6,321 grains; in Poland, the funt equals 0.894 pound.

1 KAMIAN
25.00000 (e) funts
22.55000 (a) pounds (avoirdupois)
10.2810079 (a) kilograms

The kamian is also considered to equal 22.57 pounds; in Poland, the kamian equals 22.35 pounds.

1 POOD, POUD
40.00000 (e) funts
36.08000 (a) pounds (avoirdupois)
16.4496127 (a) kilograms

The unit is also considered to equal 36.113 pounds.

1 BERKOVET,
BERKOWITZ
360.80000 (a) pounds (avoirdupois)
10.00000 (e) pood
164.4961271 (a) kilograms

The unit is also considered to equal 361.13 pounds.

1 PACKEN
1,082.40000 (a) pounds (avoirdupois)
3.00000 (e) berkovets
493.4883813 (a) kilograms

The packen is also considered to equal 1,083.4 pounds.

Weight and mass units of Spain

Spain, as is true of most countries, is currently using the metric system, but the following units persist in many measuring instances.

1 GRANO
0.770486 (a) grain
49.9266602 (a) milligrams

The grano is also used in Argentina and Chile.

1 ARIENZO,
QUILATE
4.00000 (e) granos
3.08194 (a) grains
199.7066407 (a) milligrams

The quilate in Brazil equals 3.01 grains; in Colombia, 3.09 grains; and in the Philippines, 3.16 grains.

1 TOMIN   9.24583 (a) grains
                     3.00000 (e) arienzos
                     0.5991199 (a) gram

                     The tomin is used to measure silver.

1 DINERO,   18.49167 (a) grains
ESCRUPULO   2.00000 (e) tomins
                     1.1982398 (a) grams

1 ADARME,   27.73750 (a) grains
DRACMA   3.00000 (e) tomins
                     1.7973598 (a) grams

                     The unit, as defined above, is also used in Mexico,
                     but the dracma is also considered to equal 55.4
                     grains.

1 CARACTER,   55.47500 (a) grains
OCHAVA   2.00000 (e) adarmes
                     3.5947195 (a) grams

1 CASTELLANO   71.00000 (a) grains
                     4.6007226 (a) grams

                     The castellano is used in the measure of gold.

1 ONZA   443.80000 (a) grains
                     8.00000 (e) caracters
                     28.7577563 (a) grams

                     The onza is also used in Mexico.

1 MARCO   3,550.00000 (a) grains
                     50.00000 (e) castellanos
                     230.0361305 (a) grams

1 LIBRA   2,304.00000 (e) arienzos
                     384.00000 (e) dineros
                     256.00000 (e) adarmes

1 *Libra*	128.00000 (e) caracters
(continued)	16.00000 (e) onzas
	1.01440 (a) pounds (avoirdupois)
	460.1241001 (a) grams

The libra, as defined above, is also used in Bolivia, Chile, Ecuador, Cuba, Peru, and Venezuela. But in Argentina, a libra equals 1.013 pounds; in Colombia, 1.102 pounds; in Mexico, 1.015 pounds; and in Portugal and Brazil, 1.012 pounds.

1 ARROBA	25.36000 (a) pounds (avoirdupois)
	25.00000 (e) libras
	11.5031025 (a) kilograms

The arroba, as defined above for Spain, is also used in Mexico, but in Portugal an arroba equals 32.38 (a) avoirdupois pounds or 14.6908 (a) kilograms, and in Brazil equals 32.380 (a) avoirdupois pounds or 14.6873 (a) kilograms. The term arroba also occurs under Spirits Measure and Capacity and Volume Units of Spain in Part Three.

Weight and mass units of Sweden

Sweden, as is true of most countries, is currently using the metric system, but the following units persist in many measuring instances.

1 ASS	0.741446 (a) grain
	48.0448801 (a) milligrams

1 ORT	65.60313 (a) grains
	4.2510110 (a) grams

The ort in Denmark equals 7.72 grains.

1 SKALPUND,	8,848.00000 (e) ass
PUND	6,560.31250 (a) grains
	100.00000 (e) orts
	425.1010993 (a) grams

The unit equals 0.937 pound and about 6,559 grains in Finland.

1 LISPUND,   131,206.25000 (a) grains
LISPOUND   20.00000 (e) skalpund
                        18.74375 (a) pounds (avoirdupois)
                        8.5020220 (a) kilograms

The unit is also considered to equal 131,180 grains. In Denmark the unit equals 17.64 pounds and about 123,480 grains.

1 STEN   209,930.00000 (a) grains
                        32.00000 (e) skalpund
                        29.99000 (a) pounds (avoirdupois)
                        13.6032352 (a) kilograms

1 SKEPPUND   400.00000 (e) skalpund
                        374.87500 (a) pounds (avoirdupois)
                        170.0404397 (a) kilograms

1 NYLAST   12,000.00000 (e) skalpund
                        5.62313 (a) short tons
                        5.02065 (a) long tons
                        5,101.2131911 (a) kilograms

The nylast is also considered to equal 4.686 short tons.

Weight and mass units of Yugoslavia

Yugoslavia, as is true of most countries, is currently using the metric system, but the following units persist in many measuring instances.

1 DRAMM   49.38500 (a) grains
                        3.2000942 (a) grams

1 SATLIJK	100.00000 (e) dramms
	0.705500 (a) pound (avoirdupois)
	320.0094170 (a) grams

1 OKA, OKE 400.00000 (e) dramms
4.00000 (e) satlijks
2.82200 (a) pounds (avoirdupois)
1.2800377 (a) kilograms

The unit is also used in Greece.

1 TOVAR 282.20000 (a) pounds (avoirdupois)
100.00000 (e) okas
128.0037670 (a) kilograms

The tovar in Bulgaria equals 282.6 pounds.

1 WAGON 22,046.08000 (a) pounds (avoirdupois)
9.84200 (a) long tons
10,000.0000000 (e) kilograms

Additional ancient and foreign weight and mass units

Most of the countries or territories represented below now use the metric system, but the units mentioned still persist in many measuring instances.

AUSTRIA
1 quentchen: 1/128 pfund, 67.52 grains; also in Bavaria
1 pfund: 1.235 pounds (avoirdupois)
1 saum: 275 pfunde, 339.5 pounds (avoirdupois)
1 karch: 400 pfunde, 494 pounds (avoirdupois)

COLOMBIA
1 quilate: 3.0856 grains
1 libra: 2,500 quilates, 1.102 pounds (avoirdupois)
1 saco: 125 libras, 137.75 pounds (avoirdupois)
1 carga: 250 libras, 275.5 or 275.6 pounds (avoirdupois)

INDIA

1 **dhan:** 0.4675 grain
1 **ratti, rati:** 4 dhans, 1.87 grains
1 **tola:** ⅕ chittak, 180 grains
1 **chittak, chittack:** 1/16 ser, 900 grains
1 **ser:** 14,400 grains

LIBYA

1 **kharouba:** 1/2560 rotl, 3.09 grains
1 **termino:** 1/128 rotl, 61.83 grains
1 **rotl:** 7,910.4 grains
1 **oka:** 2.831 pounds (avoirdupois)
1 **gorraf:** 9.75 okas, 27.6 pounds (avoirdupois)
1 **giarra:** 165.3 pounds (avoirdupois)

THE PHILIPPINES

1 **quilate:** 3.16 grains
1 **punto:** ⅓ catty, 0.465 pound (avoirdupois)
1 **catty:** 1.395 pounds (avoirdupois)
1 **chinanta:** 10 catties, 13.33 or 13.95 pounds (avoirdupois)
1 **fardo:** 33.5 pounds (avoirdupois); used in the measure of tobacco
1 **lachsa:** 48 catties, 66.93 pounds (avoirdupois)
1 **picul:** 139.44 pounds (avoirdupois)

TURKEY

1 **kerat:** 1/16 dirhem, 3.09 grains
1 **drachma, dram, dirhem:** 49.5 grains
1 **miskal:** 74.2 grains; in Arabia, 72 grains
1 **yusdrum, cequi:** ¼ oka, 0.707 pound (avoirdupois)
1 **oka, oke:** 2.83 pounds (avoirdupois)
1 **kileh, kile:** 20 okas, 56.6 pounds (avoirdupois)
1 **kantar:** 44 okas, 124.45 pounds (avoirdupois)
1 **cheke:** 180 okas, 509 pounds (avoirdupois); used in the measure of wood

MISCELLANEOUS FOREIGN UNITS

1 **barrel** (*cement*) (*South Africa*): 375 pounds (avoirdupois)
1 **bat, baht, tical** (*Thailand*): 15.0 grams
1 **batman** (*Iran*): 6.5 pounds (avoirdupois)
1 **bundle** (*South Africa*): 7 pounds (avoirdupois)

Miscellaneous
foreign units
(continued)

1 caja (*Costa Rica, Guatemala Nicaragua, Honduras, and El Salvador*): 35.27 pounds (avoirdupois)

1 ch'ien (*Mainland China*): 58.33 grains (silver)

1 chilogramma (*Italy*): 1/288 libra, 18.17 grains

1 daric (*ancient Persia*): 0.3 ounce, 8.4 grams (gold)

1 denaro (*Italy*): 1/288 libra, 18.17 grains

1 gran (*Bavaria*): 1/7680 pound, 1.13 grains

1 karwar (*Iran*): 100 (e) batmans, 650 pounds (avoirdupois)

1 libra (*metric*) (*Italy*): 2.205 pounds (avoirdupois)

1 livre, demikilo (*France*): 1.1 pounds (avoirdupois), 500 grams

1 mahnd (*Arabia*): 2.04 pounds (avoirdupois)

1 miskal (*Arabia*): 72.0 grains

1 miskal (*Iran*): 154.3 grains (formerly 71 grains)

1 mna (*Greece*): 1.5 kilograms, 1.172 okes

1 moosa (*Cyprus*): 112 pounds (avoirdupois)

1 ocque (*ancient Arabia*): 4 artal, 3 pounds (avoirdupois)

1 parto (*Malta*): 1/480 rotl, 25.5 grains

1 poide de marc (*France*): 0.2448 kilograms

1 ratel (*Iran*): 1.014 pounds (avoirdupois)

1 salma, salm (*Malta*): 490 pounds (avoirdupois)

1 shekel (*ancient Israel*): 0.497 ounce (avoirdupois), 14.1 grams

1 skaalpund (*Norway*): 1.098 pounds (avoirdupois)

1 tercio (*Mexico*): 160 libras, 162.35 pounds (avoirdupois)

1 tical (*ancient Siam*): 231.5 grains

1 toman, tomand (*grain*) (*Arabia*): 187.2 pounds (avoirdupois

1 tonelada (*Argentina*): 2,025.6 pounds (avoirdupois)

1 tonelada (*Brazil*): 13.5 quintals, 1,748.79 pounds (avoirdupois)

1 tonelada (*Nicaragua*): 2,028.7 pounds (avoirdupois)

1 tunna smjors (*Iceland*): 224 pund, 246.9 pounds (avoirdupois)

1 uckia (*Algeria*): 1.204 ounces (avoirdupois)

PART FIVE

The metric system

The metric system is presented in Part Five:

Unit prefixes
Length units
Comparison of metric and customary length units
Surface units
Comparison of metric and customary surface units
Capacity and volume units
Comparison of metric and customary capacity and volume units
Weight and mass units
Comparison of metric mass units and customary weight and mass units

Unit prefixes

All the following prefixes are used in the metric system and may also be used with various customary units, or with virtually any unit. Each prefix indicates that the unit name attached to it is either a part of the main unit or so many times the main unit.

Yotta-	10^{+24}	1,000,000,000,000,000,000,000,000.
Zetta-	10^{+21}	1,000,000,000,000,000,000,000.
Exa-	10^{+18}	1,000,000,000,000,000,000.
Peta-	10^{+15}	1,000,000,000,000,000.
Tera-	10^{+12}	1,000,000,000,000.
Giga-	10^{+9}	1,000,000,000.
Mega-	10^{+6}	1,000,000.
Myria-	10^{+5}	100,000.
Kilo-	10^{+3}	1,000.
Hecto-	10^{+2}	100.
Deca-	10^{+1}	10.
Deci-*	10^{-1}	0.1
Centi-	10^{-2}	0.01
Milli-	10^{-3}	0.001
Micro-	10^{-6}	0.000001
Nano-	10^{-9}	0.000000001
Pico-	10^{-12}	0.000000000001
Femto-	10^{-15}	0.000000000000001
Atto-	10^{-18}	0.000000000000000001
Zepto-	10^{-21}	0.000000000000000000001
Yocto-	10^{-24}	0.000000000000000000000001

Metric length units

1 ATTOMETER	0.001000 (e) fermi

1 FERMI,	1,000.00000 (e) attometers
FEMTOMETER	0.010000 (e) X-unit

> The fermi was named for Enrico Fermi, atomic physicist. The term fermi is also used in Linear Measure, Part One; Square Measure, Part Two; and Metric Surface Units, Part Five.

* Sometimes those units prefixed with deci-, centi-, etc., which are smaller than the main unit, are not capitalized, while the main unit itself and those larger begin with a capital letter.

---

1 X-UNIT

100.00000 (e) fermis

0.100000 (e) picometer

The X-unit is not actually a part of the metric system, but is included because of its similarity to other metric units. The term X-unit is also used in Linear Measure, Part One.

1 PICOMETER,
MICROMICRON

10.00000 (e) X-units

0.010000 (e) angstrom

1 ANGSTROM,
ANGSTROM UNIT,
TENTHMETER

100.00000 (e) picometers

0.100000 (e) nanometer

The unit is not actually a part of the metric system, but is included because of its similarity to other metric units. The terms angstrom and angstrom unit are also used in Linear Measure, Part One.

1 NANOMETER,
MILLIMICRON,
BICRON

10.00000 (e) angstroms

0.001000 (e) micrometer

The term millimicron is also used in Linear Measure, Part One.

1 MICROMETER,
MICRON

1,000.00000 (e) nanometers

0.001000 (e) millimeter

The term micron is also used in Linear Measure, Part One.

1 MILLIMETER

1,000.00000 (e) micrometers

0.100000 (e) centimeter

The millimeter is known as a streep in the Netherlands, a royal gramme in Greece, a mou in Persia, and a strich in Germany. The strich in Switzerland, however, equals 3.0 (e) millimeters.

1 CENTIMETER

0.100000 (e) decimeter

0.010000 (e) meter

The centimeter is equal to or known as a duim in the Netherlands, a royal dakylos in Greece, a kung

1 Centimeter *(continued)*	fen in Mainland China, a khat in Turkey, and a dito in Italy.

1 DECIMETER

10.00000 (e) centimeters

0.100000 (e) meter

The decimeter is also known as a palm, which equals 1/10 el, in the Netherlands. The term palm is also used in Linear Measure, Part One.

The decimeter is also known as a palame in Greece and a palmo in Italy; however, the palmo in Portugal and Brazil equals 8.66 inches, and in Spain, 8.23 inches.

1 METER

10.00000 (e) decimeters

0.100000 (e) decameter

J.B.J. Delambre, a French astronomer and mathematician, and his colleague M. Mechain, estimated the length of the meridian from the equator to the pole by the measurement of an arc between Dunkirk and Barcelona. The ten-millionth part of this meridian was taken as the standard unit of length of their system and was termed a meter.

The meter was later determined to equal 1,553,164.3 wavelengths of red cadmium light, and in 1960, a meter equal to 1,650,763.73 wavelengths of orange light of isotopic krypton 86 at 760 millimeters pressure and 15 degrees centigrade was adopted as the new international standard of length. The meter is now considered to equal the length of the path traveled by light in vacuum during a time interval of 1/299792458 of a second. The name meter was derived from the Greek word *metron,* meaning "a measure."

The meter is also known as a stab in Germany, an el in the Netherlands, a ken in Siam, a kung ch'ih in China, a zira and zirai in Turkey, and an arsmin in new Turkey (an arshin in old Turkey equaled 75.79 centimeters and 288 hatt). A decameter (10 meters) is known as a kette in Germany and a roede, which equals 10 ellen, in the Netherlands. A hectometer (100 meters) is known as a yin or kung in China.

1 KILOMETER	1,000.00000 (e) meters
	0.010000 (e) myriameter

The kilometer closely approximates one ten-thousandth of the earth's circumference from pole to pole. It is also known as a kung li in China, a miglio in Italy, a mijl in the Netherlands, and a stadion in modern Greece. The term stadion also occurs under Linear Measure in Part One.

The nymil of Sweden, the mil of modern Sweden (the old mil equaled 6.64 statute miles), and the farsang of Persia equal 10.0 (e) kilometers.

1 MYRIAMETER	100.00000 (e) kilometers
	0.100000 (e) megameter
1 MEGAMETER	1,000.00000 (e) kilometers
	0.001000 (e) gigameter
1 GIGAMETER	1,000.00000 (e) megameters
	0.001000 (e) terameter
1 TERAMETER	1,000,000,000.00000 (e) kilometers
	1,000.00000 (e) gigameters

Comparison of metric and customary length units

The following material shows the relation between various metric and customary length units. To identify the section in which a particular unit appears, the following code is used:

(M) Metric Length Units
(L) Linear Measure, Part One
(S) Solar Measure, Part One

1 CENTIMETER **(M)**	0.393701 (r) inch (L)

1 METER (M)	3.28084 (r) feet (L) 1.09361 (r) yards (L) 0.198838 (r) rod (L)
1 HECTOMETER (M)	0.497096 (r) furlong (L)
1 KILOMETER (M)	0.621371 (r) statute mile (L)
1 TERAMETER (M)	0.105700 (r) light-year (S)

Metric surface units

1 MILLIBARN

10^{-31} (e) centare
0.001000 (e) barn

The millibarn is not actually a part of the metric system, but is included because of its similarity to other metric units. The term millibarn is also used in Square Measure, Part Two.

1 BARN

10^{-28} (e) centare
1,000.00000 (e) millibarns

The barn is not actually a part of the metric system, but is included because of its similarity to other metric units. The term barn is also used in Square Measure, Part Two.

1 FERMI, SQUARE FEMTOMETER

10^{-17} (e) centare

The fermi was named for Enrico Fermi, atomic physicist. The term fermi is also used in Linear Measure, Part One; Square Measure, Part Two; and Metric Length Units, Part Five.

1 SQUARE MILLIMETER

0.010000 (e) square centimeter
0.000001 (e) centare

1 SQUARE CENTIMETER	100.00000 (e) square millimeters 0.000100 (e) centare
1 SQUARE DECIMETER	100.00000 (e) square centimeters 0.010000 (e) centare
1 MILLARE	10.00000 (e) square decimeters 0.100000 (e) centare
1 CENTARE, SQUARE METER	10.00000 (e) millares 0.010000 (e) are
1 DECIARE	10.00000 (e) centares 0.100000 (e) are
1 ARE, SQUARE DECAMETER	100.00000 (e) centares 10.00000 (e) deciares 0.000100 (e) square kilometer

The are equals and is known as the kung mu in Mainland China, a square kafiz in Persia, and a vierkante roede in the Netherlands.

1 DECARE   10.00000 (e) ares
0.100000 (e) hectare

The decare is also known as a royal stremma in Greece.

1 HECTARE   10,000.00000 (e) centares
100.00000 (e) ares
0.010000 (e) square kilometer

The hectare equals and is known as a djerib in Turkey, a jerib in Persia, a kung ch'ing in Mainland China, a manzana in Argentina, and a bunder in the Netherlands.

1 KILIARE   1,000.00000 (e) ares
0.100000 (e) square kilometer

1 SQUARE	1,000,000.00000 (e) centares
KILOMETER,	10,000.00000 (e) ares
MYRIARE	100.00000 (e) hectares

Comparison of metric and customary surface units

The following material shows the relation between various metric and customary surface units. To identify the section in which a particular unit appears, the following code is used:

(**M**) Metric Surface Units
(**S**) Square Measure, Part Two

1 SQUARE CENTIMETER (M)	0.155000 (r) square inch (S)
1 MILLARE (M)	1.07639 (r) square feet (S)
1 CENTARE (M)	1.19599 (r) square yards (S)
1 ARE (M)	0.024711 (r) acre (S)
1 SQUARE KILOMETER (M)	0.386102 (r) section (S)
	0.010725 (r) township (S)

Metric capacity and volume units

The following code is used to differentiate between a capacity unit and a volume unit:

(**C**) Capacity Unit
(**V**) Volume Unit

Originally all metric units were based upon the meter. However, it was found that a higher degree of accuracy could be attained with masses than with volumes, and it was therefore deemed preferable to use a specifically defined standard of mass than one derived from the unit of length through the unit of volume.

A defined mass, the international prototype kilogram, was therefore adopted as the standard of mass, and the unit of volume, the liter, was redefined in terms of the mass standard—that is, as the volume of a kilogram of pure water at the temperature of its maximum density. The standard value of the liter was thereby corrected from 1.000027 to 1.000028 cubic decimeters.

1 CUBIC
MILLIMETER (V)

0.00100000 (e) cubic centimeter (V)

0.00099997 (r) milliliter (C)

1 MICROLITER
(C)

0.00000100 (e) liter (C)

Generally, the prefixes micro, centi, deci, deca, kilo, hecto, and myria, are not used with the unit liter. They are included here, however, so that the user can better visualize the relative sizes in capacity and volume measure.

1 CUBIC
CENTIMETER (V),
CENTIMETER
CUBE (V)

1,000.00000000 (e) cubic millimeters (V)

0.99997200 (r) milliliter (C)

0.00100000 (e) cubic decimeter (V)

1 MILLILITER (C)

1.00002800 (r) cubic centimeters (V)

0.10000000 (e) centiliter (C)

1 CENTILITER (C)

10.00028000 (r) cubic centimeters (V)

10.00000000 (e) milliliters (C)

1 DECILITER (C)

10.00000000 (e) centiliters (C)

0.10000280 (r) cubic decimeter (V)

1 CUBIC
DECIMETER (V),
MILLISTERE (V)

1,000.00000000 (e) cubic centimeters (V)

9.99972001 (r) deciliters (C)

0.10000000 (e) centistere (V)

1 LITER (C)   10.00000000 (e) deciliters (C)
1.00002800 (r) cubic decimeters (V)
0.10000000 (e) decaliter (C)

1 CENTISTERE (V)   10.00000000 (e) cubic decimeters (V)
9.99972001 (r) liters (C)
0.10000000 (e) decistere (V)

1 DECALITER (C)   10.00000000 (e) liters (C)
1.00002800 (r) centisteres (V)
0.10000000 (e) hectoliter (C)

The decaliter is known as a schepel in the Netherlands.

1 DECISTERE (V)   10.00000000 (e) centisteres (V)
9.99972001 (r) decaliters (C)
0.10000000 (e) stere (V)

1 HECTOLITER (C)   10.00000000 (e) decaliters (C)
1.00002800 (r) decisteres (V)
0.10000000 (e) kiloliter (C)

The hectoliter is also known in the Netherlands as a vat, mud, and zak; a zak also equals 2.838 bushels.

1 STERE (V), CUBIC METER (V)   10.00000000 (e) decisteres (V)
9.99972001 (r) hectoliters (C)
0.10000000 (e) decastere (V)

The stere is used in measuring firewood and is known as a wisse in the Netherlands.

1 KILOLITER (C)   10.00000000 (e) hectoliters (C)
1.00002800 (r) steres (V)
0.10000280 (r) decastere (V)

1 DECASTERE (V)   10.00000000 (e) steres (V)
9.99972001 (r) kiloliters (C)
0.10000000 (e) hectostere (V)

1 MYRIALITER	10.00028000 (r) steres (V)
(C)	10.00000000 (e) kiloliters (C)
1 HECTOSTERE	10.00000000 (e) decasteres (V)
(V)	0.10000000 (e) kilostere (V)
1 KILOSTERE (V),	10.00000000 (e) hectosteres (V)
CUBIC	0.00100000 (e) cubic hectometer (V)
DECAMETER (V)	
1 CUBIC	1,000.00000000 (e) kilosteres (V)
HECTOMETER (V)	0.00100000 (e) cubic kilometer (V)
1 CUBIC	
KILOMETER (V)	1,000.00000000 (e) cubic hectometers (V)

Comparison of metric and customary capacity and volume units

The following material shows the relation between various metric and customary units of capacity and volume. To identify the section in which a particular unit appears, the following code is used:

(**MV**) Metric Volume Unit
(**MC**) Metric Capacity Unit
(**C**) Cubic Measure, Part Three
(**D**) Dry Measure, Part Three
(**L**) Liquid Measure, Part Three
(**F**) Apothecaries' Fluid Measure, Part Three
(**S**) Spirits Measure, Part Three

1 MILLILITER	0.270520 (r) fluid dram (F)
(MC)	
1 LITER (MC)	33.81497 (r) fluid ounces (F)
	2.11344 (r) pints (L)

1 Liter (mc)	1.75975 (r) British pints (F)
(continued)	1.32090 (r) fifths (S)
	0.879882 (r) imperial quart (L)

1 CUBIC — 61.02361 (r) cubic inches (C)
DECIMETER (MV) — 0.908058 (r) quart (D)

1 DECALITER (MC) — 2.64180 (r) gallons (L)

1 CENTISTERE (MV) — 1.13507 (r) pecks (D)

1 DECISTERE (MV) — 3.53131 (r) cubic feet (C)
2.83768 (r) bushels (D)

1 KILOLITER (MC) — 8.38665 (r) barrels (L)

1 STERE (MV) — 8.64817 (r) barrels (D)
1.30789 (r) cubic yards (C)

Metric weight and mass units

1 MICROGRAM — 0.000001 (e) gram

1 MILLIGRAM — 0.100000 (e) centigram

1 DECIGRAM — 10.00000 (e) centigrams
0.100000 (e) gram

1 GRAM — 10.00000 (e) decigrams
0.100000 (e) decagram

The gram is known as a wichje in the Netherlands. The gram has been defined as the mass of one cubic centimeter (a cube that is 1/100 meter on each side) of water at its temperature of maximum density.

1 DECAGRAM   10.00000 (e) grams
                       0.100000 (e) hectogram

                       The decagram equals and is known as a miscal in
                       Iran, a unit that in old Iran equaled 71 grains.

1 HECTOGRAM   10.00000 (e) decagrams
                       0.100000 (e) kilogram

1 KILOGRAM,   10.00000 (e) hectograms
KILO   0.100000 (e) myriagram

                       In Germany, Austria, and Switzerland, a kilogram is
                       known as a zollpfund and a metric pfund. The metric
                       hundredweight equals 50.0 (e) kilograms.

1 MYRIAGRAM   10.00000 (e) kilograms
                       0.100000 (e) quintal

1 QUINTAL,   10.00000 (e) myriagrams
QUINTAL   0.100000 (e) metric ton
METRIQUE,
METRIC QUINTAL,
METRIC
CENTNER,
DOUBLE
CENTNER

1 METRIC TON,   10.00000 (e) quintals
TONNEAU,
MILLIER

Comparison of metric mass units and customary weight and mass units

The following material shows the relation between
various metric and customary units of weight and
mass. To identify the section in which a particular
unit appears, the following code is used:

(**M**) Metric Unit
(**A**) Avoirdupois Weight, Part Four
(**T**) Troy Weight, Part Four
(**P**) Apothecaries' Weight, Part Four

1 GRAM (M)

0.771618 (r) scruple (P)
0.643015 (r) pennyweight (T)
0.564384 (r) dram (A)
0.257206 (r) dram (P)
0.034285 (r) assay ton (T)

1 HECTOGRAM (M)

3.52387 (r) ounces (A)
3.21508 (r) ounces (T)

1 KILOGRAM (M)

2.67923 (r) pounds (T)
2.20462 (r) pounds (A)

The kilogram described above equals 15,432.356 grains, but in Britain, the kilogram equals 15,432.3564 grains and 2.2046223 avoirdupois pounds. The U.S. Post Office generally equates 15 grams with 0.5 ounce.

1 QUINTAL (M)

0.055116 (r) last (A)

PART SIX

Diverse units

Diverse units dealing primarily with count and the senses are presented in Part Six:

Units of count

Paper measure

Units of book size

Time measure

Sound measure

Units used in music

Units of poetic verse

Light measure

Units of count

1 DUAD, BINARY, 2 units
DOUBLE, DUAL,
BINAL,
TWOFOLD,
DUPLEX

1 TRIAD, TRIPLE, 3 units
TRIPLEX,
TERNARY

1 TETRAD, 4 units
QUADRUPLET,
QUADRUPLE,
QUADRUPLEX,
QUATERNION,
QUARTERNARY

1 QUINTET, 5 units
QUINTUPLE,
QUINTUPLET

1 HEXAD, 6 units
SENARY

1 HEBDOMAD, 7 units
HEPTAD, SEPTET,
SEPTUPLE

1 OCTUPLE, 8 units
OCTAD, OCTAVE,
OCTONARY,
OGDOAD

1 ENNEAD, 9 units
NONUPLE

1 DENARY	10 units
1 DOZEN, TWELVEMO, DUODECIMO	12 units The term duodecimo is also used in Units of Book Size, Part Six.
1 BAKER'S DOZEN, LONG DOZEN	13 units The baker's dozen started in England in the time of Cromwell, when bakers began making rolls smaller to save flour; when customers became angry, the bakers added an extra roll.
1 SEPTEN-DECIMAL	17 units
1 SCORE	20 units The term score also occurs under Avoirdupois Weight in Part Four.
1 SEXAGENARY	60 units
1 HUNDRED, CENTUPLE	100 units
1 GROSS	12 dozen or 144 units
1 THOUSAND	1,000 units
1 GREAT GROSS	12 gross or 1,728 units
1 MILLION	1,000,000 units
1 BILLION	1,000,000,000 units In Britain, the billion is known as a milliard.
1 TRILLION	1 (followed by 12 zeros) units The British trillion, however, equals 1 followed by 18 zeros units, equal to the U.S. quintillion.

---

1 QUADRILLION   1 (followed by 15 zeros) units

The British quadrillion, however, equals 1 followed by 24 zeros units, equal to the U.S. and French septillion.

1 QUINTILLION   1 (followed by 18 zeros) units

The British quintillion, however, equals 1 followed by 30 zeros units, equal to the U.S. nonillion.

1 SEXTILLION   1 (followed by 21 zeros) units

The sextillion of Britain and Germany, however, equals 1 followed by 36 zeros units, equal to the U.S. undecillion. The sextillion defined above is also used in France.

1 SEPTILLION   1 (followed by 24 zeros) units

The septillion of Britain and Germany, however, equals 1 followed by 42 zeros units, equal to the U.S. tredecillion. The septillion defined above is also used in France.

1 OCTILLION   1 (followed by 27 zeros) units

The octillion of Britain and Germany, however, equals 1 followed by 48 zeros units, equal to the U.S. quindecillion. The octillion defined above is also used in France.

1 NONILLION   1 (followed by 30 zeros) units

The British nonillion, however, equals 1 followed by 54 zeros units, equal to the U.S. septendecillion.

1 DECILLION   1 (followed by 33 zeros) units

The decillion of Britain and Germany, however, equals 1 followed by 60 zeros units, equal to the U.S. novemdecillion. The decillion defined above is also used in France.

1 UNDECILLION   1 (followed by 36 zeros) units

1 DUODECILLION   1 (followed by 39 zeros) units

---

1 TREDECILLION	1 (followed by 42 zeros) units
1 QUATTUOR-DECILLION	1 (followed by 45 zeros) units
1 QUINDECILLION	1 (followed by 48 zeros) units

The quindecillion is also known as the Buddhist asanka.

1 SEXDECILLION	1 (followed by 51 zeros) units
1 SEPTEN-DECILLION	1 (followed by 54 zeros) units
1 OCTO-DECILLION	1 (followed by 57 zeros) units
1 NOVEM-DECILLION	1 (followed by 60 zeros) units
1 VIGINTILLION	1 (followed by 63 zeros) units
1 GOOGOL	1 (followed by 100 zeros) units

The googol is also known as a googolplex, but the googolplex is more often considered to equal 10 to the power of 1 googol.

1 QUINTO-QUADAGIN-TILLION	1 (followed by 138 zeros) units

1 ASANKHYEYA The Buddhist asankhyeya equals 10^{+140} units or 100 quinto-quadagintillions.

1 MILLI-MILLIMILLION The unit, devised by Rudolf Ondrejka, equals 10 raised to the power of 6 billion.

1 SKEW Named for Stanley Skewes, professor at Cape Town University, South Africa, the unit equals 10 to the power 10 to the power 10 to the power 3.

Paper measure

Paper measure concerns the units used in measuring and counting all paper products made for newspaper, book, and magazine printing and for stationery.

1 QUIRE	24.00000 (e) sheets (or units)
1 COMMERCIAL QUIRE, PRINTER'S QUIRE	25.00000 (e) sheets 1.04167 (r) quires
1 REAM, SHORT REAM	480.00000 (e) sheets 20.00000 (e) quires
1 COMMERCIAL REAM	500.00000 (e) sheets 20.00000 (e) commercial quires
1 PERFECT REAM	516.00000 (e) sheets 1.07500 (e) reams
1 BUNDLE	960.00000 (e) sheets 2.00000 (e) reams
1 COMMERCIAL BUNDLE	1,000.00000 (e) sheets 40.00000 (e) commercial quires 2.00000 (e) commercial reams
1 PRINTER'S BUNDLE	1,920.00000 (e) sheets 4.00000 (e) reams
1 BALE	4,800.00000 (e) sheets 5.00000 (e) bundles

The term bale also occurs under Avoirdupois Weight in Part Four.

1 COMMERCIAL BALE	5,000.00000 (e) sheets 5.00000 (e) commercial bundles
1 PRINTER'S REAM	5,160.00000 (e) sheets

Units of book size

In the binding of books, magazines, pamphlets, and similar publications, various standardized sizes exist in the printing industry.

1 FOLIO	A sheet of paper folded once to make 2 leaves or 4 pages of a book.
1 ATLAS FOLIO	The largest book-size folio, with leaves 16 by 25 inches.
1 QUARTO	A volume printed from sheets folded twice to form 4 leaves or 8 pages, the book size being about 9.5 by 12 inches.
1 SIXMO	The sixmo indicates the size of a piece of paper cut six from a sheet.
1 OCTAVO	A book size determined by printing on sheets folded to form 8 leaves or 16 pages, of a size about 6 by 9 inches. The unfolded octavo must be more than 20 but less than 25 centimeters in height and varies as follows:

cap octavo: 4.25 by 7 inches

crown octavo: 5 by 7.5 inches

demy octavo: 5.5 by 8 inches

imperial octavo: 8.25 by 11.5 inches

medium octavo: 6 by 9.5 inches

post octavo: 5.5 by 7.5 inches

royal octavo: 6.5 by 10 inches

1 DUODECIMO, TWELVEMO   A book size, about 5 by 7.5 or 7.75 inches, determined by printing on sheets folded to form 12 leaves or 24 pages.

1 SEXTODECIMO, SIXTEENMO   A volume printed from sheets folded to form 16 leaves or 32 pages, approximately 4 by 6 inches.

1 OCTODECIMO, EIGHTEENMO   A book size, about 4 by 6.25 inches, determined by printing on sheets folded to form 18 leaves or 36 pages.

1 THIRTYTWOMO   A book size, about 3.5 by 5.5 inches, determined by printing on sheets folded to form 32 leaves or 64 sheets.

1 DEMY   The demy indicates a size of paper, which in the United States is 16 by 21 inches and in Great Britain is 17.5 by 22 inches.

1 HEXATEUCH   A collection of six books, usually the first six books of the Old Testament.

1 HEPTATEUCH   A collection of seven books, usually the first seven books of the Old Testament.

1 OCTATEUCH   A collection of eight books, usually the first eight books of the Old Testament.

Time measure

Time measure is based on the relation between the earth and the moon or sun. However, the speed of the earth's rotation about its axis varies slightly, and these variations affect time measure. They are classified as:

Secular variation. Tidal friction acts as a brake and causes a secular increase in the length of the day—about 1.0 millisecond per century.

Irregular variation. The speed of rotation may increase for 5 to 10 years, then decrease, with the maximum difference from the mean length of the day during a century being about 5 milliseconds. Accumulated difference in time has amounted to about 40 seconds since A.D. 1900, with the cause probably being turbulent motion in the core of the earth.

Periodic variation. Seasonal variations exist, with periods lasting 1 year and other periods lasting 6 months. The probable cause of this annual variation is seasonal change in wind patterns between the north and south hemispheres and semiannual variation due to tidal action of the sun, which distorts the shape of the earth.

The cumulative effect of these three types of variations is such that each year the earth is late about 30 milliseconds near June 1 and ahead about 30 milliseconds near October 1. The maximum seasonal variation in the length of the day is about 0.5 millisecond.

Furthermore, the earth has been slowing down. In the late Cretaceous period, some 85,000,000 years ago, the earth rotated so much faster than it does today that a year consisted of 370.3 days, and in Cambrian times, some 600,000,000 years ago, a year equaled about 425 days. The earth has "lost" 11 seconds just since 1958, compared with the atomic clock, which does not deviate a second in 20,000 to 60,000 years, or even longer.

To correct our time system, which is *universal coordinate time*, 1.0 second was added to our clocks at precisely 4:59 and 60 seconds mountain standard time on December 31, 1973, and similarly throughout the world. This marked the third time in 2 years the leap second had been added to coordinate the world's atomic clocks with the earth's rotation.

Such variations in the earth's behavior have made it difficult to construct workable calendars

throughout the ages. Calendars, whose name originated with the Roman *calendae* or calends, "the
first day of the month," have generally been of
three types: lunar, solar, and lunisolar

The *lunar calendar* is based on the 29.5-day
lunar or synodic month and is related to the move-
ments of the moon. The Muhammadan calendar is
an example of this type. A lunar year equals
354.367056 days; this is 354 days with an added
leap period of 11.012 days every 30 years.

The *solar calendar's* accuracy depends on the two
equinoxes and two solstices occurring on or about
the same days each year. A solar year equals
365.2422 days. Examples of this type calendar are
the Gregorian improvement of the Julian calendar,
which is primarily our present system, and the old
Mayan and Egyptian calendars.

The *lunisolar calendar* assumes that the lunar
month and the tropical year work in harmony,
with a thirteenth lunar month being added every
2 or 3 years. The lunisolar year is only 354 (e)
days long. The Jewish calendar is an example of
this type.

The *Julian calendar* began in 46 B.C., when
Julius Caesar extended the year 46 B.C. by adding
23 days to Februarius and two new months, of 33
and 34 days, between November and December,
so that 46 B.C. suddenly contained 445 days instead
of the previous 354+ days. He then decreed that
succeeding years would equal 365 days and 6
hours.

This was indeed an improvement, but the system still presented problems because the Julian
year exceeded the true solar year by 11 minutes
and 14 seconds. The centuries passed until finally
Aloysius Lilius, a Neapolitan astronomer, proposed
to Pope Gregory XIII on February 24, 1582, that
10 days be dropped from the 1582 calendar year;

hence October 4 of that year was suddenly succeeded by October 15.

Then, the Julian calendar was further corrected by instituting the leap year every fourth year instead of every third year, with only the century years divisible by 400 to be leap years.

This became known as the *Gregorian calendar*, in use today, but not adopted by England until 1752 and by Russia until 1917.

Although we presently use the Gregorian calendar system, other improved systems have been proposed. Probably the most practical new system is *decimal time*, mentioned in the Introduction. Other meritorious proposals are the *world calendar*, where the "quarter year" is the basic unit, and the *perpetual calendar*, devised by Willard E. Edwards of Hawaii and once introduced in the U.S. House of Representatives. In both systems the year comprises 8 months of 30 days each and 4 months of 31 days each, with a "world day" at the end of the year and a "leap day" every fourth year.

The *international fixed calendar* was designed by Moses Bruines Cotsworth (1859–1943), an English statistician, and is a revised version of the calendar proposed by Auguste Comte (1798–1857), a French philosopher. The year in this case comprises 13 months (the seventh named Sol) of 28 days each with a "year day" at the end of the year and a "leap day" every fourth year.

Still another proposed system is based on the moon and involves the lunation; days are called *lunes* and hours *lunours*. Individual units of this system are defined in this section.

But the Gregorian calendar is with us now, and so that time and dates can be considered in their proper relation, an understanding of *time zones* is important. Originally, time was determined in relation to the sun, eventually with the use of the sun dial. Called *sun time*, this system is still used in many rural or remote areas and by nature's wildlife itself, but it varies from place to place. Noon in the western hemisphere is nighttime and darkness in Europe. The sun may be said to "travel" about 15 degrees of longitude per hour.

In the United States, since 1883, and at the request of railroads, which apparently tired of setting up different time schedules for every town, *standard time* has been in use, dividing the country into four continental zones plus three Alaska standard zones (covering the 135th, 150th, and 165th meridians west of Greenwich). There are twenty-four zones worldwide.

In the continental United States, the zones are eastern standard (75 degrees), central standard (90 degrees), mountain standard (105 degrees), and western or pacific standard (120 degrees west of Greenwich). The clocks are 1 hour later in each successive belt from east to west, and there is a 1-day difference as the international date line is crossed.

Even with the zone system, however, the sun at high noon may not coincide with noon on your clock. One influencing factor is that each zone is about 1,037.6 statute miles wide at the equator, and the sun cannot be exactly perpendicular to more than one point along such a distance. A second factor is that the earth does not move in a circle around the sun (which would make its speed uniform); it moves in an ellipse, appearing to go faster on the short sides. Several times a year it is as much as 14.5 minutes slow or fast (a circumstance that accounted for some very complicated sun dials and bad watches before 1883). Standard noon, therefore, is not "true noon" but "mean noon," an average of where the sun ought to be if its "speed" were constant.

The most accurate timekeeping devices now existing are the twin atomic hydrogen masers installed in 1964 at the U.S. Naval Research Laboratory in Washington, D.C. They are based on the frequency of the hydrogen atom's transition period of 1,420,450,751,694 cycles per second, allowing accuracy to within 1.0 second per 1,700,000 years. The most accurate, and most complicated, clock in the world is probably the Olsen clock in Copenhagen, Denmark, which has more than 14,000 units and is accurate to 0.5 second in 300 years. The celestial pole motion of the Olsen clock will take 25,700 years to complete a full circle.

1 CHRONON

The chronon equals about one-billionth of a trillionth of a second. Particles of time related to the activity of the electron within an atom, and important to the cesium beam clock, which indicates that the outer single electron of the cesium atom, when it changes one position to another, either emits or absorbs energy at the rate of 9,192,631,770 vibrations per second.

1 PICOSECOND 10^{-12} (e) second

1 NANOSECOND 10^{-9} (e) second

1 MICROSECOND   0.000001 (e) second

1 MILLISECOND   1,000.00000 (e) microseconds
0.001000 (e) second

1 SECOND   1,000.00000 (e) milliseconds
0.016667 (r) minute

The second is also known as an ephemeris second and is defined as 1/31,556,925.9747 tropical year 1900.0 at 12 hours ephemeris time. It was further defined in 1967, by the thirteenth General Conference on Weights and Measure, as the duration of 9,192,-631,770 ± 20 particular oscillations within a cesium 133 atom and termed an atomic second.

The term second also occurs under Angular Measure in Part Two.

1 CENTILUNOUR   35.43671 (r) seconds
0.100000 (e) decilunour
0.010000 (e) lunour

1 MINUTE   60.00000 (e) seconds
0.016667 (r) hour

The term minute also occurs under Angular Measure in Part Two.

1 DECILUNOUR   354.36706 (r) seconds
10.00000 (e) centilunours
0.100000 (e) lunour

1 QUARTER HOUR   900.00000 (e) seconds

1 LUNOUR   3,543.67056 (r) seconds
100.00000 (e) centilunours
10.00000 (e) decilunours
0.041667 (r) lune
0.041015 (r) lunar mean solar day
0.001389 (r) lunation

1 SIDEREAL HOUR   3,590.17080 (r) seconds
59.83618 (r) minutes
0.997270 (r) hour
0.041667 (r) sidereal day

1 HOUR   3,600.00000 (e) seconds
60.00000 (e) minutes
1.00274 (r) sidereal hours
0.041780 (r) sidereal day
0.041667 (r) common day

1 LUNE   85,048.09333 (r) seconds
24.00000 (e) lunours
0.984353 (r) lunar mean solar day
0.033333 (r) lunation

1 SIDEREAL DAY, 86,164.09100 (r) seconds
EXACT DAY 1,436.06833 (r) minutes
24.00000 (e) sidereal hours
23.93447 (r) hours
0.997270 (r) common day
0.994546 (r) mean solar day

Sidereal denotes time determined by the position of the stars. The sidereal day equals 23 hours, 56 minutes, and 4.091 seconds. The day is lengthening 0.00117 millisecond per day or 0.0427 second per century.

1 COMMON DAY 86,400.00000 (e) seconds
1,440.00000 (e) minutes
24.00000 (e) hours
1.00274 (r) sidereal days
0.997270 (r) mean solar day
0.142857 (r) week

A calendar day is that period from 1.0 second after midnight to the following midnight.

1 MEAN SOLAR 86,636.55500 (r) seconds
DAY 1,443.94258 (r) minutes
24.06571 (r) hours
1.00548 (r) sidereal days
1.00274 (r) common days

The mean solar day equals 24 hours, 3 minutes, and 56.555 seconds. A true solar day or apparent day equals one complete rotation of the earth in relation to the sun, but this varies, and the average is the mean solar day.

1 LUNAR DAY 89,400.00000 (e) seconds
1,490.00000 (e) minutes
24.83333 (r) hours

The lunar day equals 24 hours and 50 minutes.

1 QUARTAN	4.00000 (e) common days

The quartan is also considered to equal 72 hours. The terms biweekly and semiweekly indicate twice a week or sometimes every 2 weeks.

1 WEEK,	10,080.00000 (e) minutes
HEBDOMAD	168.00000 (e) hours
	7.00000 (e) common days
	0.500000 (e) fortnight

1 FORTNIGHT	336.00000 (e) hours
	14.00000 (e) common days
	2.00000 (e) weeks

1 NODICAL	2,351,135.80000 (r) seconds
MONTH,	27.21222 (r) common days
DRACONTIC	0.995997 (r) tropical month
MONTH	0.987576 (r) anomalistic month
	0.921101 (r) synodic month
	0.907074 (r) common month

The nodical month is defined as the mean time of revolution of the moon from ascending node to ascending node, and equals 27 common days, 5 hours, 5 minutes, and 35.8 seconds.

1 TROPICAL	2,360,584.70000 (r) seconds
MONTH	27.31324 (r) common days
	1.00402 (r) nodical months
	0.991545 (r) anomalistic month
	0.925196 (r) synodic month
	0.910719 (r) common month

The tropical month is defined as the mean time of revolution of the moon from any point of the ecliptic (the great circle of the celestial sphere that is the apparent path of the sun among the stars or of the earth as seen from the sun) back to the same point, and equals 27 common days, 7 hours, 43 minutes, and 4.7 seconds. However, 1/12 tropical year equals 30 common days, 10 hours, 29 minutes, and 3.8 seconds—or 2,629,743.7925 seconds.

1 SIDEREAL MONTH	2,360,591.50000 (r) seconds
	27.39646 (r) sidereal days
	27.31332 (r) common days

The sidereal month is nearly synonymous with the tropical month. It is defined as the mean time of revolution of the moon from any star back to the same star, and equals 27 common days, 7 hours, 43 minutes, and 11.5 seconds.

1 ANOMALISTIC MONTH

2,380,713.10000 (r) seconds
27.55455 (r) common days
1.01258 (r) nodical months
1.00853 (r) tropical months
0.933085 (r) synodic month
0.918485 (r) common month

The anomalistic month is defined as the mean time of revolution of the moon from perigee to perigee, and equals 27 common days, 13 hours, 18 minutes, and 33.1 seconds.

1 LUNAR CIVIL MONTH, LUNAR MONTH

2,548,800.00000 (e) seconds
0.983333 (r) common month

1 SYNODIC MONTH

2,551,442.80000 (r) seconds
29.53059 (r) common days
1.08519 (r) nodical months
1.08085 (r) tropical months
1.07171 (r) anomalistic months
0.984353 (r) common month

The synodic month equals 29 common days, 12 hours, 44 minutes, and 2.8 seconds. It is also known as a lunar synodic month, which equals 2,551,442.83333 seconds.

The synodic month is also known as a lunation, the unit of lunar or moon time that is based on the orientation of the moon to the sun and that averages 29.530589 terrestrial mean solar days and equals 720 lunours and 30 lunes.

1 COMMON	2,592,000.00000 (e) seconds
MONTH, MONTH	720.00000 (e) hours
	30.00000 (e) common days
	4.28571 (r) weeks
	1.10245 (r) nodical months
	1.09803 (r) tropical months
	1.08875 (r) anomalistic months
	1.01695 (r) lunar civil months

The common month is also known as a lunar month, but a lunar month is also considered to equal 28 common days or 4 weeks.

A calendar month equals 1/12 calendar year and is also known as a solar month.

A metonic cycle equals 235 lunar months and about 19 Julian years, at the end of which times the phases of the moon recur in the same order and on the same days as in the preceding cycle.

1 BIMESTRIAL	2.00000 (e) common months

1 QUARTER	7,889,231.37750 (r) seconds
	0.250000 (e) tropical year

Biquarterly indicates twice every 3 common months. The quarter is also known as a trimester, which equals 3 common months.

1 ANOMALISTIC	330.65460 (r) common days
YEAR	12.00000 (e) anomalistic months

The anomalistic year is the average interval between consecutive passages of the earth through the perihelion.

1 COMMON	30,585,600.00000 (e) seconds
LUNAR YEAR,	354.00000 (e) common days
LUNAR YEAR	12.00000 (e) lunar civil months
	0.998967 (r) lunar astronomical year
	0.969220 (r) tropical year

1 TROPICAL YEAR, ASTRONOMICAL YEAR, COMMON YEAR, NATURAL YEAR, YEAR, EQUINOCTIAL YEAR	31,556,925.51000 (r) seconds 365.24219 (r) common days 52.17746 (r) weeks 1.03176 (r) common lunar years 1.03069 (r) lunar astronomical years 0.999961 (r) sidereal year

Biyearly and semiyearly indicate twice a year, or sometimes every 2 years. Biannual means twice a year, while biennial means every 2 years.

The tropical year, meaning a year of the seasons, is also known as a cosmic year. This is the time of rotation of our galaxy from the sun's distance from the center, or about 200,000,000 to 225,000,000 years.

The tropical year is also known as a solar year, which equals 12 solar months, and as an ephemeris year, which was defined in 1956 as the time of one earth revolution around the sun and is equal to 31,556,925.9747 seconds.

The tropical year equals 365 common days, 5 hours, 48 minutes, and is decreasing at the rate of 0.530 second each century.

1 SIDEREAL YEAR, ASTRAL YEAR, EXACT YEAR	31,558,149.54000 (r) seconds 13.36874 (r) sidereal months 1.00004 (r) tropical years

The sidereal year is defined as the time in which the sun's center, deporting eastward from the ecliptic meridian of a given star, returns to the same point; it equals 365 common days, 6 hours, 9 minutes, and 9.54 seconds.

A calendar year, or Julian year, based on 7 calendar months of 31 common days, 4 calendar months of 30 common days, and 1 calendar month of 28 days, equals 365 common days. Because the calendar year more closely approximates 365 common days, 5 hours, 49 minutes, and 12 seconds of mean solar time, a leap year of 366 common days replaces the calendar year every fourth year to take up the slack of the unprecise Gregorian calendar. The calendar year is also known as a civil year and a legal year; a fiscal year equals a calendar year, but may differ in point of beginning.

1 TRIENNIUM, TRIENNIAL	3.00000 (e) common years
1 QUADRENNIUM, QUADRENNIAL	4.00000 (e) common years
1 QUIN-QUENNIUM, QUINQUENNIAL, PENTAD	5.00000 (e) common years
1 SEXENNIAL	6.00000 (e) common years
1 SEPTENNIUM, SEPTENARY, SEPTENNATE, SEPTENNIAL, HEPTAD	7.00000 (e) common years
1 OCTENNIUM, OCTENNIAL	8.00000 (e) common years
1 DECADE, DECENNIUM, DECENNARY	10.00000 (e) common years 0.100000 (e) century 0.010000 (e) millennium
1 QUIN-DECENNIUM, QUINDECENNIAL	15.00000 (e) common years
1 VICENNIAL	20.00000 (e) common years
1 SEPTUAGENARY	70.00000 (e) common years
1 CENTURY, CENTENARY	100.00000 (e) common years 10.00000 (e) decades 0.100000 (e) millennium

1 BICENTENARY, BICENTENNIAL	200.00000 (e) common years
1 TERCENTENARY, TRICENTENNIAL	300.00000 (e) common years
1 QUADRI- CENTENNIAL	400.00000 (e) common years
1 SEXCENTENARY	600.00000 (e) common years
1 SEPT- CENTENARY	700.00000 (e) common years
1 MILLENNIUM, MILLIAD	1,000.00000 (e) common years 100.00000 (e) decades 10.00000 (e) centuries
1 KALPA	The kalpa, in the Hindu calendar, equals about 4,320 million years.

Sound measure

Sound travels at the rate of 1,142 feet, or 3/14 statute mile, per second, or 1 statute mile in about 4.67 seconds. In water, sound travels at the rate of 4,708 feet per second.

Sound velocity is affected by temperature as well as by the medium through which it travels. The increase in the velocity of sound is roughly 0.5 meter per second for each degree centigrade rise in temperature. At 20 degrees centigrade, sound velocity is 344 meters per second, and at 100 degrees, 386 meters per second.

1 DECIBEL,   0.115128 (r) neper
TRANSMISSION   0.100000 (e) bel
UNIT

The decibel is the usual unit for measuring the loudness of sounds and equals the loss in power over a statute mile of standard cable at 860 cycles. On a decibel scale, a sound 20 decibels higher than another is 100 times greater in intensity; a sound 30 decibels higher is 1,000 times greater in intensity.

The decibel scale is an arbitrary scale usually extending between the high and low tolerance of the human ear and is guided by the sound pressure level, which equals 20 log (Pe/Po)db. The reference sound pressure, Po, usually taken as the standard, is 0.0002 microbar (or about 1/1,000,000 atmosphere) and represents the low threshold of hearing.

One of the loudest noises ever created in a laboratory was 210 decibels, reported by the National Aeronautics and Space Administration in 1965. The highest note yet attained is one of 60,000 megahertz (see Electrical Frequency Units, Part Eight) generated by a laser beam at Massachusetts Institute of Technology, Cambridge, Massachusetts, in 1964.

1 NEPER   8.686000 (a) decibels

1 BEL   10.000000 (e) decibels

1.151278 (a) nepers

The number of bels can be either the common log of a power ratio or twice the common log of a current ratio (2 bels is a power ratio of 100, since the log of 100 is 2, and 2 bels is also a current ratio of 10, since the log of 10 is 1). The bel was named for Alexander Graham Bell.

1 FREQUENCY   The frequency represents the number of oscillations of an atomic radiating source in a unit of time. In harmonic motions, it is the number of vibrations or cycles in a unit of time (example: 25 or 60 cycles per second).

1 MILLIOCTAVE   0.001000 (e) octave

1 OCTAVE   1,000.000000 (e) millioctaves

In a vibration series, the octave is an interval analogous to the musical octave; the wave number at its beginning and end are to each other as 1:2; an interval embracing 8 diatonic degrees. The term octave also occurs under Units Used in Music and Units of Poetic Verse in Part Six and under Spirits Measure in Part Three.

1 MEL   The mel is a unit of pitch; how high or low a note sounds to the listener. A tone vibrating at a frequency of 1,000 vibrations per second has a pitch of 1,000 mels. But 200 on the frequency scale equals 300 mels. A frequency of 1,900 equals 1,500 mels, a frequency of 4,000 equals 2,250 mels, and a frequency of 14,000 equals 3,250 mels.

Units used in music

Music, like virtually everything man encounters or devises, has its own group of measuring aids. The organized scientific study of such units is referred to as musicology.

Music is an agreeable combination of sound and time elements, encompassing the time length of a certain sound, its tonal quality, its loudness, and even the quiet periods between such sounds. To complement an understanding of the musical units themselves, the following definitions of other musical terms may be helpful.

A *rest* is an interval of silence between tones, while a *caesura* is a pause or breathing at a point of rhythmic division in a melody.

A *measure* is a group of beats or pulses made by the regular recurrence of primary, or heavy, accents. An *octave* is a tone on the eighth degree from a given tone counted as the first. A *second* is a tone on the next degree from a given tone or the interval between such tones.

Major denotes an interval greater by a half step than the corresponding minor interval, and a *minor* is an interval less by a half step than the corresponding major interval.

A *semiditone* or *hemiditone* is a minor third in Greek music, and a *sextuple* is a unit characterized by having six beats or pulses to the measure.

1 HEMIDEMISEMI-QUAVER 0.500000 (e) demisemiquaver

The unit is equivalent to one sixty-fourth whole note.

1 DEMISEMI-QUAVER 2.00000 (e) hemidemisemiquavers
0.500000 (e) semiquaver

The unit is equivalent to one thirty-second whole note.

1 SEMIQUAVER 4.00000 (e) hemidemisemiquavers
2.00000 (e) demisemiquavers
0.500000 (e) quaver

The unit is equivalent to one sixteenth whole note.

1 QUAVER 2.00000 (e) semiquavers
0.125000 (e) semibreve

The unit is equivalent to one eighth whole note.

1 CROTCHET, QUARTER TONE 2.00000 (e) quavers
0.500000 (e) minim

The unit is an interval equivalent to half a semitone, a quarter note.

1 MICROTONE The microtone is an interval smaller than a half tone.

1 MINIM, SEMITONE, HALF TONE, HALF STEP, HALF NOTE	2.00000 (e) crotchets 0.500000 (e) semibreve The minim was originally the smallest interval used in music, but as music expanded and became more sophisticated, extending its range in modern times, the partial quavers were named to cover shorter time ranges. The hemidemisemiquaver is now the smallest musical interval used.
1 SEMIBREVE, WHOLE NOTE	2.00000 (e) minims
1 BREVE	2.00000 (e) semibreves The breve is the longest modern note. An allabreve is an expression denoting a species of time in which every measure contains a breve, or four minims. It is a time value also of two minims or four crotchets to a measure, but taken at a rate twice as fast.
1 TRIAD	The triad is a chord of three tones, especially one consisting of a given root tone with its major or minor third and its perfect augmented or diminished fifth. A tritone is an interval composed of three whole tones; a tercet or triplet is a group of three notes performed in the time of two of the same value.
1 TETRACHORD	The tetrachord is a diatonic series of four tones, the first and last separated by a perfect fourth.
1 PENTATONE	The pentatone is an ancient and medieval interval of five whole steps, an augmented sixth.
1 HEXACHORD	In medieval music the hexachord is a diatonic series of six tones, with a half step between the third and fourth tones and whole steps between the others.
1 SEPTIMOLE, SEPTAVE	The unit is a group of seven notes to be played in the time of four or six of the same value.

Units of poetic verse

Poetic verse, which consists of words coordinated to produce a flowing and, usually, a rhyming quality, has its own group of measuring aids. Analyzing the length of a group of words, the number of lines of such words, and where and how certain words are repeated or rhymed with others necessitates a degree of organized measure.

To complement an understanding of such poetic measuring units, the following definitions of other terms used in poetry may be helpful.

A *stanza* or *stave* is a group of lines of verse, commonly four or more in number, arranged and repeated according to a fixed plan as regards the number of lines, the meter, and the rhyme, and forming a poem.

Caesura indicates a pause in a line of verse dictated by sense or natural speech rhythm rather than by metrics. *Strophe* means (1) in the Greek choral dance, the movement of the chorus while turning from one side to the other of the orchestra, and (2) the strain, or part of the Greek choral ode, sung during the strophe of the dance. A mono-strophe is a poem in which all the strophes or stanzas are of the same metric form.

An *envoy* is a postscript to a poem, essay, or book; specifically, a short stanza appended to a ballade and some other metrical forms.

A *sestina* is a poem of six six-line stanzas and a three-line envoy, originally without rhyme, in which each stanza repeats the end words of the lines of the first stanza, but in different order, and the envoy uses the six words again, three in the middle of the lines and three at the end.

A *terza rima* is a verse form consisting of a series of triplets having ten-syllable or eleven-syllable lines of which the middle line of one triplet rhymes with the first and third lines of the following triplet. A *triseme* is a syllable or foot of three morae. A mora is a common short syllable.

An *arsis* was formerly considered to be the unaccented syllable of a foot in verse; it is now the unstressed part of a rhythmic unit.

1 METRICAL FOOT, MEASURE

The metrical foot may be construed as a syllable. The term metrical means (1) pertaining to or characterized by versification or measure in music or poetry, and (2) composed of or making up a unit of poetic meter.

The metrical foot may have more than one syllable. A spondee is a metrical foot consisting of two long or stressed syllables. An anapest is a metrical foot consisting of three syllables, the first two short or unaccented, the last one long or accented. An amphimacer is a metrical foot, also of three syllables, but the middle syllable is short and the others are long.

The dactyl, in accentual verse, is a metrical foot consisting of one accented syllable followed by two unaccented; in quantitative verse it is one long syllable followed by two short. The word *dactyl* derives from the Middle English *dactile*, the Latin *dactylus*, and the Greek *daktylos*, or the three joints of the finger.

The iamb is a foot consisting of a short syllable followed by a long one, or of an unaccented syllable followed by an accented one. A choriamb is defined as (1) a metrical foot consisting of a trochee followed by an iamb, much employed in aeolic poetry and in the choric odes of tragedy; and (2) a foot of verse used in lyric poetry having two unstressed syllables flanked by two rhythmic stresses marking the first and last syllables of the foot.

A tetragram is a word of four letters. The trochee is a metrical foot of two syllables, a long followed by a short, or an accented followed by an unaccented. Feminine rhyme refers to double rhyme in verses with unstressed final syllables; a rhyme of two syllables of which the second is unstressed.

1 DIPODY, DIPODIE, DIMETER

A line of verse having two metrical feet.

1 TRIMETER, TRIPODY	A line of verse containing three metrical feet.
1 TETRAMETER	The tetrameter as used in ancient poetry consisted of four dipodies (eight feet) in trochaic, iambic, or anapestic meter. It now has four measures. The tetraseme is a tetrasemic foot, such as a tetrabrach (word or foot of four short syllables) or a spondee.
1 PENTAMETER	The pentameter is a line of verse consisting of five metrical feet. The unit may also be a verse consisting of two dactyls, a long syllable, two or more dactyls, and another single syllable.
1 HEXAMETER, SENARIUS	The hexameter is a dactylic verse of six feet used in Greek and Latin epic and other poetry. The first four feet are dactyls or spondees, the fifth is ordinarily a dactyl, and the last is a trochee or spondee with a caesura usually following the long syllable in the third foot.
1 SEPTENARIUS, HEPTAMETER	The unit is a verse of seven feet or stresses, often printed in two lines, especially in Greek or Medieval Latin verse, the trochaic tetrameter catalectic, or in Middle English poetry. It may also be an iambic verse of seven and a half feet.
1 OCTAMETER, OCTASTICH	The unit is a line of verse containing eight metrical feet.
1 HEMISTICH	The hemistich is an exact or approximate half stich of poetic verse.
1 LINE, STICH, VERSE	The unit is a unit of verse made up of a row of words, or such a row formed of a certain number of metrical feet characteristic of the verse. A monostich is a poem or epigram consisting of a single metrical line.
1 DISTICH	The distich is a group of two lines of verse, usually making complete sense; a rhyming couplet.

1 TRISTICH, TRIPLET	The unit is a group of three rhyming lines. A tercet is a group of three lines rhyming together or connected by rhymes with the adjacent group or groups of three lines.
1 QUATRAIN, TETRASTICH	The quatrain is a stanza or poem of four lines, usually with alternate rhymes; either of the halves of a sonnet. The tetrastich is a stanza, epigram, or poem consisting of four lines or verses.
	The pantun or pantoum is a Malay verse form, usually four lines, the third rhyming with the first and the fourth with the second.
1 PENTASTICH	The pentastich is a strophe, stanza, or poem consisting of five lines or verses.
	The rondelet is a short poem of fixed form, consisting of five lines on two rhymes, and having the opening word or words used after the second and fifth lines as an unrhymed refrain.
1 HEXASTICH, HEXASTICHON	The unit is a strophe, stanza, or poem consisting of six lines or verses. A sestet indicates the last six lines of a sonnet.
1 HEPTASTICH	The heptastich is a strophe, stanza, or poem consisting of seven lines or verses.
1 OCTAVE, OCTET	The unit is a group or stanza of eight lines, such as the first eight lines of a sonnet (see below).
	A triolet is a poem or stanza of eight lines constructed on two rhymes in which the first line is repeated as the fourth and seventh, and the second as the eighth, the scheme being *abaaabab*.
	An ottavarima is an Italian stanza of eight lines, each of eleven syllables (or in the English adaptation, of ten or eleven syllables), the first six lines rhyming alternately and the last two forming a couplet with a different rhyme.
1 RONDEAU	The rondeau is a short poem, of French origin, of fixed form, consisting of ten or thirteen lines on two rhymes, and having the opening words or phrases used in two places as an unrhymed refrain.
1 RONDEL, SONNET	The rondel is a short poem of fixed form, consisting usually of fourteen lines on two rhymes, of which four are made up of the initial couplet repeated in

1 Rondel sonnet *(continued)*	the middle and at the end; the second line of the couplet sometimes is omitted at the end.

The sonnet is a fixed verse form, of Italian origin, consisting of fourteen lines, typically five-foot iambics, and generally of two standard types: (1) the Italian sonnet, regular sonnet, or Petrarchan sonnet, in which the lines are grouped into an octave of two quatrains running on two rhymes, as *abba abba*, and a sestet (the last six lines) of two tercets on two or three rhymes, as *cdc dcd* or *cde cde;* and (2) the English sonnet, Elizabethan sonnet, or Shakespearean sonnet, in which the lines are grouped into three quatrains and a couplet, as *abab cdcd efef gg.* The Spenserian sonnet unites the quatrains by interlacing the rhymes, as *abab bcbc cdcd ee.*

Light measure

This section deals with units of photometry, which is the measurement of the properties of light, especially of luminous intensity.

Presently, the most powerful light in the world is in the light at Creac'h d'Ouessant on Île d'Ouessant, Finisterre, Brittany, France; it has a luminous intensity of up to 490,500,000 candles.

The brightest steady artificial light sources are laser beams, whose luminosities exceed the sun's 1,500,000 candles per square inch by a factor of over 1,000. The longest sustained source is a 200-kilowatt high-pressure xenon-arc lamp of 600,000 candle power, reported from the U.S.S.R. as continuously burning since 1965. In 1969, the U.S.S.R. announced blast waves traveling through a luminous plasma of inert gases heated to 90,000 degrees Kelvin; the flare-up for up to 3.0 microseconds shone at 50,000 times the brightness of the sun, or about 40,000,000,000 candles per square inch.

The most powerful searchlight developed was produced in World War II by General Electric

Company, Ltd., England; it had a consumption of 600 kilowatts and gave an arc luminance of 300,-000 candles per square inch and a maximum beam intensity of 2,700,000,000 candles from its 10-foot-diameter parabolic mirror.

1 MICROLUX	0.001000 (e) millilux
1 LUMEN PER SQUARE CENTIMETER	100.00000 (e) microluxes 0.000009 (r) foot-candle
1 MILLILUX	0.001000 (e) lux 0.000093 (r) foot-candle
1 MICROPHOT	10.00000 (e) milliluxes 0.000929 (r) foot-candle 0.000001 (e) phot
1 LUX, CANDLE-METER, LUMEN PER SQUARE METER	1,000.00000 (e) milliluxes 100.00000 (e) microphots 0.092903 (r) foot-candle The lux, the international unit of illumination, is the direct illumination on a surface that is everywhere 1.0 meter from a uniform point source of 1.0 international candle.
1 MILLIPHOT	10.00000 (e) luxes 0.929030 (r) foot-candle 0.001000 (e) phot
1 FOOT-CANDLE, CANDLE-FOOT, LUMEN PER SQUARE FOOT	10.76391 (r) luxes 0.001076 (r) phot The unit is equal to the amount of direct light thrown by 1.0 international candle on a square foot of surface every part of which is 1.0 foot away.

---

1 PHOT

10,000.00000 (e) luxes

1,000.00000 (e) milliphots

929.03040 (r) foot-candles

The phot is the cgs (centimeter-gram-second) unit of illumination. It is defined as the direct illumination on a surface that is everywhere 1.0 centimeter from a uniform point source of 1.0 international foot-candle. The cgs subsystem of the metric system is usually used for mechanics, electricity, and magnetism, but also encompasses a photometric unit.

1 LUMEN

0.079577 (r) candlepower

The lumen, the international system unit of luminous flux, is the light emitted in a unit solid angle (steradian) by a uniform point source of 1.0 international candle or $\frac{1}{4}\pi$ candlepower.

1 HEFNER CANDLE

The official unit of light intensity in Germany is the Hefner candle, which equals 0.9 candle. It is the light from the flame of a specially constructed lamp invented by von Hefner-Alteneck, German physicist, which burns amyl acetate.

1 CANDLE, CANDELA, CANDLEPOWER

12.56640 (r) lumens

The candle is a unit of luminous intensity approximately equal to the intensity of light from a $\frac{7}{8}$-inch sperm candle burning at the rate of 120 grains per hour. The candle originally equaled 1.02 candelas. The candela is also known as a standard candle, which also equals 1/60 of the luminous intensity per square centimeter of a black body radiating at the temperature of solidification of platinum (2,046 degrees Kelvin). The candela is now considered to be the luminous intensity, in a given direction, of a source that emits monochromatic radiation of frequency 540 x 10^{+12} hertz (Hz) and that has a radiant intensity in that direction of 1/683 watt per steradian.

The candle-hour is a unit of light equal to the total luminous energy emitted in 1.0 hour by a source having a luminous intensity of 1.0 candle.

An international candle, British candle, or British standard candle, equals the light given off by the flame of a sperm candle $\frac{7}{8}$-inch in diameter burning at the rate of 7.776 grams per hour, which approximates 120 grains per hour. It is also defined as the light emitted by 5.0 square millimeters of platinum at the temperature of solidification.

A violle or violle's standard, devised by Jules Violle, French physicist, is a photometric unit defined as the light emitted by a square centimeter of platinum at the temperature of solidification. The unit is about equal to 18.5 British standard candles.

The bougie decimale or decimal candle is a French photometric standard having the value of 1/20 violle platinum standard, or slightly less than a British candle.

1 CANDLE-LUMEN

The candle lumen is a unit of flux of light, the $\frac{1}{4}\pi$ part of the total flux of light emitted by a source having a mean spherical intensity of 1.0 candlepower.

1 KILOLUMEN

1,000.00000 (e) lumens

1 CANDLE PER SQUARE CENTIMETER

0.155000 (r) candle per square inch

1 CANDLE PER SQUARE INCH

6.45160 (e) candles per square centimeter
0.006944 (r) candela per square foot

1 FOOT-LAMBERT

45.83943 (r) candles per square inch
0.318309 (r) candela per square foot
3.426259 (r) candela per square meter

The foot-lambert is a unit of luminescence equal to $1/\pi$ candela per square foot.

1 CANDELA PER SQUARE FOOT

929.03040 (e) candles per square centimeter
144.00000 (e) candles per square inch
3.14159 (r) foot-lamberts

1 CANDELA PER SQUARE METER

10.76391 (r) candelas per square foot

1 MILLILAMBERT

0.001000 (e) lambert

1 LAMBERT

1,000.00000 (e) millilamberts
3,183.099000 (r) candela per square meter

The lambert is a unit of brightness equal to the light of a surface which is radiating or reflecting 1.0 lumen per square centimeter. A perfect diffusing surface emitting 1.0 lumen per square foot has a brightness

1 Lambert
(continued)

of 1.076391 millilamberts. The lambert was named for Johann Lambert (1728–1777), German physicist.

1 PHOTON

The photon is a quantum of light energy, the energy being proportional to the frequency of the radiation.

PART SEVEN

Other
diverse units

Additional diverse units, mostly of an engineering nature, are presented in Part Seven:

Pressure units

Temperature units

Energy units

Force units

Radiation units

Viscosity units

Units expressing flow of water

Speed units

Miscellaneous units

Additional units defining various capacities and volumes per various masses and weights, various lengths per various capacities and volumes, various masses and weights for various areas, and various capacities and volumes per various areas.

Pressure units

One of the highest sustained laboratory pressures ever reported is about 5,000,000 atmospheres, or 36,700 tons per square inch, achieved in the U.S.S.R. about 1958. Momentary pressure of 75,000,000 atmospheres, or 550,000 tons per square inch, was achieved in 1958 in the United States using dynamic methods and impact speed of up to 18,000 miles per hour.

One of the highest laboratory vacuums achieved is about 1.0×10^{-16} atmosphere, while pressure in interstellar space has been estimated at 1.0×10^{-19} atmosphere. At sea level there are $3 \times 10^{+19}$ molecules per cubic centimeter in the atmosphere, but in interstellar space there are probably less than 10 molecules per cubic centimeter.

1 MICROBAR, DYNE PER SQUARE CENTIMETER
0.100000 (e) newton per square meter
0.001000 (e) millibar

1 NEWTON PER SQUARE METER, PASCAL
10.00000 (e) microbars
0.101972 (r) kilogram per square meter
0.020885 (r) pound per square foot

1 KILOGRAM PER SQUARE METER
9.80665 (r) newtons per square meter
The term kilogram per square meter is also used in Miscellaneous Units, Part Seven.

1 POUND PER SQUARE FOOT
47.88194 (r) newtons per square meter
0.006944 (r) pound per square inch

1 MILLIBAR
1,000.00000 (e) microbars
100.00000 (e) newtons per square meter
0.100000 (e) centibar
0.014503 (r) pound per square inch

1 TORR   133.322400 (r) pascal

1 KILOPASCAL   1,000.00000 (e) newtons per square meter
                          0.009869 (r) atmosphere

1 POUND PER   6,895.00000 (a) newtons per square meter
SQUARE INCH   144.00000 (e) pounds per square foot
                          0.070262 (r) kilogram per square centimeter
                          0.068000 (r) atmosphere

The pound per square inch is based on the avoirdupois pound and is also considered to equal 0.070307 kilogram per square centimeter.

1 KILOGRAM PER 98,066.50000 (a) newtons per square meter
SQUARE   14.22284 (r) pounds per square inch
CENTIMETER   0.980665 (r) bar
                    0.967841 (r) atmosphere

The kilogram per square centimeter is also known as the metric atmosphere, which equals 14.2234 pounds per square inch and is defined as the pressure of 1.0 kilogram per square centimeter or that of a column of mercury 735.514 millimeters in height.

1 BAR,   100,000.00000 (e) newtons per square meter
MEGADYNE PER   100.00000 (e) centibars
SQUARE   14.50326 (r) pounds per square inch
CENTIMETER   1.01972 (r) kilograms per square centimeter
                    0.986970 (a) standard atmosphere
                    0.001000 (e) kilobar

The bar nearly equals the pressure of 29.5306 inches or 750.076 millimeters of mercury at 0 degree centigrade and under standard conditions of gravity.

1 ATMOSPHERE   2,116.10232 (r) pounds per square foot
                    1,013.23066 (r) millibars
                    14.69516 (r) pounds per square inch
                    1.03321 (r) kilograms per square centimeter
                    1.01323 (r) bars

1 Atmosphere
(continued)

The atmosphere is defined as the standard pressure under which the mercury barometer stands at 760 millimeters when the density of the mercury is at 0 degree centigrade and the acceleration of gravity has the value of 980.665 on the cgs (centimeter-gram-second) system. More simply, it is the pressure of air at sea level, and also equals $1.01325 \times 10^{+5}$ newtons per square meter and 2,116.8 pounds per square foot.

Steam, under ordinary circumstances, is equal to the pressure of the atmosphere. For example, 1 cubic inch of water is converted into about 1 cubic foot of steam, producing a force equal to the effort needed to raise 2,200 pounds to a height of 1 foot. Its weight is 0.488 that of the air, and it takes 27,206 cubic feet of steam to equal 1 avoirdupois pound.

A cubic foot of boiler will heat 2,000 cubic feet of space to an average of about 70 to 80 degrees Fahrenheit, and 1 square foot of steam pipe is adequate to warm 200 cubic feet of space.

And some additional information of interest: a cubic foot of air at the earth's surface weighs 1.222 ounces; 13.27 cubic feet weigh 1 pound.

1 STANDARD ATMOSPHERE

2,116.35360 (a) pounds per square foot
14.69690 (a) pounds per square inch

1 HECTOBAR

100.00000 (e) bars
0.647490 (r) long ton per square inch

1 LONG TON PER SQUARE INCH

2,240.00000 (e) pounds per square inch
152.32000 (a) atmospheres
1.54443 (r) hectobars

1 KILOBAR

14,503.26200 (a) pounds per square inch
1,000.00000 (e) bars

Temperature units

Thermometry is the measurement of temperature, and there are four chief systems of temperature measure:

The *Fahrenheit scale*, named for Gabriel Daniel Fahrenheit (1686–1736), a German physicist, is based on the melting point of ice (32 degrees above zero) and the boiling point of water (212 degrees above zero).

The *Celsius or centigrade scale*, devised in 1742 by Anders Celsius (1701–1744), Swedish astronomer, set 100 degrees as the freezing point and 0 degrees as the boiling point of pure water; this was later reversed to 0 degree as the freezing point and 100 degrees as the boiling point. To convert to the Fahrenheit scale, merely multiply degrees centigrade by 9/5 and add 32 degrees.

The *Kelvin scale*, devised by William Thomson, Lord Kelvin (1824–1907), a British physicist and mathematician, and also known as the Rankine scale after William Rankine, is based on thermodynamic principles in which zero is equivalent to −459.4 degrees Fahrenheit or −273 degrees centigrade.

The *Reaumur, Reaumer, or Reaum scale*, introduced in Spain by R.A.F. de Reaumur (1683–1757), French naturalist, sets the freezing point of water at 0 degree and the boiling point at 80 degrees.

Some of the highest man-made temperatures, those produced in the center of a thermonuclear fusion bomb, reach 300,000,000 to 400,000,000 degrees centigrade. One of the highest laboratory temperatures, reported by the U.S.S.R. in 1969, was 50,000,000 degrees centigrade for 1/50 second.

One of the lowest temperatures ever reached is 5×10^{-7} degrees Kelvin, achieved by the Centre d'Etudes Nucléaires, Saclay, France, in 1969. On the Kelvin scale absolute zero is 273.15 degrees centigrade or −459.67 degrees Fahrenheit.

**1 SMALL
CALORIE,
CALORIE, GRAM
CALORIE, GRAM
DEGREE, MEAN
CALORIE, THERM-
OCHEMICAL
CALORIE**

0.010000 (e) centuple calorie
0.003968 (r) btu
0.001000 (e) large calorie

The unit is defined as the amount of heat required, at a pressure of 1.0 atmosphere, to raise the temperature of 1.0 gram of water 1.0 degree centigrade, and is equivalent to 4.186 joules, an energy unit.
The steam calorie is equivalent to 4.1868 joules and equals 0.003972 (r) btu.

**1 CENTUPLE
CALORIE,
RATIONAL
CALORIE**

100.00000 (e) small calories
0.396825 (r) btu
0.100000 (e) large calorie

The unit is also equivalent to 418.571 joules.

**1 BTU, BRITISH
THERMAL UNIT**

252.00000 (e) small calories
2.52000 (e) centuple calories
0.252000 (e) large calorie

The btu is defined as the quantity of heat required to raise the temperature of 1.0 avoirdupois pound of water 1.0 degree Fahrenheit at or near its point of maximum density (39.1 degree Fahrenheit), and is equivalent to 1,054.8 joules.
A centigrade thermal unit is defined as the heat required to raise the temperature of 1.0 avoirdupois pound of water 1.0 degree centigrade.

**1 LARGE
CALORIE, GREAT
CALORIE,
KILOGRAM
CALORIE,
KILOCALORIE**

1,000.00000 (e) small calories
10.00000 (e) centuple calories
3.96825 (r) btus

The unit is the amount of heat required to raise 1.0 kilogram of water 1.0 degree centigrade, and is equivalent to 4,185.714 joules.
A degree day, used for estimating quantities of fuel and power consumption, is determined by subtracting the mean temperature of a calendar day from 65 degrees Fahrenheit.

**1 KILOGRAM
CALORIE PER
SECOND**

0.066138 (r) btu per minute

1 BTU PER MINUTE	15.11944 (r) kilogram calories per second
1 BTU PER SQUARE FOOT	0.271245 (r) gram calorie per square centimeter
1 GRAM CALORIE PER SQUARE CENTIMETER	3.68667 (r) btu per square foot

Energy units

1 MILLI-EQUIVALENT	0.000001 (e) electron volt
1 ELECTRON VOLT, EQUIVALENT VOLT	1.60207×10^{-19} (r) joule The electron volt is equal to the energy acquired by an electron falling through a potential difference of 1.0 volt.
1 LANGLEY	The langley is a value of energy per unit area equal to 1 calorie per square centimeter, and is the electromagnetic radiation incident upon a surface.
1 MEV	1,000,000.00000 (e) electron volts The letters of the unit stand for million electron volts.
1 ERG, DYNE-CENTIMETER	10^{-7} (e) joule 0.000074 (r) foot-pound The erg, the name derived from the German word *ergon* meaning "work," is defined as the work done when a force of 1.0 dyne is applied to an object, moving it a distance of 1.0 centimeter and is equivalent to 9.48×10^{-11} btu and 2.778×10^{-11} watt hour.
1 MICROJOULE	10.00000 (e) ergs

1 GRAM-
CENTIMETER
  0.000098 (r) joule

1 BEV
  1,000.00000 (e) mevs

  The letters of the unit stand for billion electron volts.

1 OUNCE-INCH
  7.06155 (r) newton-millimeters
  0.167595 (r) foot-poundals

1 NEWTON-
MILLIMETER
  0.141612 (r) ounce-inch

1 FOOT-POUNDAL 421,641.00000 (a) ergs
  0.421402 (r) joule
  0.031080 (r) foot-pound

  The foot-poundal, a unit of work, is defined as the work done when a force of 1.0 poundal acts through a distance of 1.0 foot, with an acceleration of gravity of 32.16, expressed in feet per second per second. The foot-poundal is also considered to equal 421,641 ergs.

1 METER-GRAM,
GRAM-METER
  0.009812 (r) joule

  The unit is equal to the work done in lifting a weight of 1.0 gram through a vertical distance of 1.0 meter.

1 INCH-POUND
  16.00000 (r) ounce-inches
  2.68153 (r) foot-poundals
  0.113000 (r) joule
  0.083333 (r) foot-pound

  The inch-pound is based on the avoirdupois pound.

1 JOULE, $62,419,245 \times 10^{+11}$ (a) electron volts
WATT-SECOND, 10,000,000.00000 (e) ergs
NEWTON-METER, 23.73031 (r) foot-poundals
ABSOLUTE JOULE 0.737463 (r) foot-pound
  0.101919 (r) kilogram-meter

The joule is the international system unit of energy, equal to the work done when a current of 1.0 ampere is passed through a resistance of 1.0 ohm for 1.0 second. It is also equal to the work done when the point of application of a force of 1.0 newton is displaced 1.0 meter in the direction of the force.

The joule was named for James Prescott Joule (1818–1889), English physicist, and is equivalent to 0.238892 small calorie and 0.0009486 btu, both heat units.

A joule-meter is an integrating watt-meter for measuring the energy in joules expended in an electrical circuit or developed by a machine.

1 INTERNATIONAL JOULE

1.00017 (r) joules

1 FOOT-POUND

13,559,974.56000 (a) ergs

32.16000 (r) foot-poundals

12.00000 (e) inch-pounds

1.35600 (r) joules

0.000377 (r) watt-hour

The foot-pound is equal to the work done in raising 1.0 avoirdupois pound against the force of gravity the height of 1.0 foot, and is also a unit of moment equal to the moment of a force of 1.0 avoirdupois pound acting at a distance of 1.0 foot from the center of moments. The foot-pound is also considered to equal $1.3825 \times 10^{+7}$ ergs, and is equivalent to 0.001286 btu, a heat unit.

1 KILOGRAM-METER

9.81170 (a) joules

7.23577 (a) foot-pounds

1 MANPOWER

The unit describes the average man's ability to lift a weight of 90 avoirdupois pounds 1.0 foot in 1.0 second. Also, it is a unit assumed to be equal to the rate at which a man can do mechanical work (raise 70 pounds 1.0 foot high in 1 second for 10 hours a day) and is taken as 1/10 horsepower. Although sometimes considered an energy unit, horsepower is listed in this book under Electrical Power Units in Part Eight, along with other variations of the horsepower unit.

1 INCH-TON   226.00045 (a) joules

166.66667 (r) foot-pounds

The inch-ton is equal to the work done in raising 1.0 short ton against the force of gravity the height of 1.0 inch.

1 FOOT-TON   3,037.44608 (a) joules

13.44000 (e) inch tons

The foot-ton is a unit of work equivalent to the energy expended in raising a long ton of 2,240 avoirdupois pounds to a height of 1.0 foot.

1 WATT-HOUR   3,600.00000 (e) joules

2,654.86726 (r) foot-pounds

1 HORSEPOWER-   2,684,879.98020 (r) joules

HOUR   1,980,000.00000 (e) foot-pounds

0.745800 (r) kilowatt-hour

0.187500 (e) ton-mile

The horsepower-hour is equivalent to 2,544.986607 btus, a heat unit.

1 KILOWATT-   1,000.00000 (e) watt-hours

HOUR, KELVIN,   1.34071 (r) horsepower-hours

BOARD OF   0.404432 (r) ton-kilometer

TRADE

COMMERCIAL   The kilowatt-hour is equivalent to 3,412.088985 btus,

UNIT   a heat unit.

1 TON-   8,901,363.20642 (r) joules

KILOMETER   6,561,677.76000 (r) foot-pounds

3.31504 (r) horsepower-hours

2.47260 (r) kilowatt-hours

0.621371 (r) ton-mile

The ton-kilometer is based on a short ton of 2,000 avoirdupois pounds.

1 TON-MILE   14,325,359.89440 (r) joules

10,560,000.00000 (e) foot-pounds

5.33333 (r) horsepower-hours

3.97927 (r) kilowatt-hours

1.60934 (r) ton-kilometers

The ton-mile is based on the short ton of 2,000 avoirdupois pounds and the statute mile of 5,280 feet.

1 THERM (U.S.)   105.480400 (r) megajoules

1 HORSEPOWER-   8,765.81264 (r) horsepower-hours

YEAR   6,538.19672 (r) kilowatt-hours

1,643.59090 (r) ton-miles

The horsepower-year is based on the tropical year of 31,556,925.51 seconds.

Force units

1 DYNE   0.001020 (r) gram

0.001000 (e) kilodyne

0.000072 (r) poundal

The dyne is a force under whose influence a body with 1.0-gram mass experiences an acceleration of 1.0 centimeter per second per second. The term dyne also occurs under Troy Weight in Part Four.

1 GRAM   980.39216 (r) dynes

0.070814 (r) poundal

The term gram also occurs under Metric Weight and Mass Units in Part Five.

1 KILODYNE,   1,000.00000 (e) dynes

CENTINEWTON   1.02000 (a) grams

0.010000 (e) newton

1 POUNDAL   13,826.00000 (a) dynes

14.10000 (a) grams

| *1 Poundal* | 0.138260 (r) newton |
| *(continued)* | 0.031080 (r) pound |

The poundal is a unit of force that imparts to a mass of 1.0 avoirdupois pound an acceleration equal to 1.0 foot per second per second. It corresponds to the dyne, except that pound and foot replace gram and centimeter.

1 OUNCE-FORCE 0.278014 (r) newton

1 NEWTON

100.00000 (e) centinewtons

7.23300 (r) poundals

0.224800 (r) pound

The newton was named for Sir Isaac Newton (1642–1727), English mathematician, scientist, and philosopher. It is the force acting for 1.0 second upon 1.0 kilogram of mass, that imparts to it a velocity of 1.0 meter per second.

1 POUND,
POUND FORCE

453.59237 (e) grams

32.17400 (a) poundals

16.00000 (e) ounce forces

4.44823 (r) newtons

The pound and pound-force are units of force that impart to a mass of 1.0 avoirdupois pound an acceleration equal to 1.0 foot per second per second. The pound is also a British unit of force equal to the weight of a standard 1.0 pound mass where the local acceleration of gravity is 32.174 feet per second; and the pound-force is also defined as the gravitational weight of a 1.0 pound mass at sea level and standard latitude (45 degrees).
The term pound also occurs under Troy Weight, Avoirdupois Weight, and Apothecaries' Weight in Part Four.

1 KILOGRAM
FORCE,
KILOPOND

9.80665 (e) newtons

1 MEGADYNE

1,000.00000 (e) kilodynes

0.100000 (e) dyne-seven

1 DYNE-SEVEN   10,000,000.00000 (e) dynes

1 SLUG,   The slug is a British engineering system unit de-
GEEPOUND,   scribed as 1.0 pound of force acting for 1.0 second
METRIC SLUG   upon 1.0 unit of mass and imparting to it a velocity
of 1.0 foot per second. A slug is also that unit of mass
which when acted upon by a force of 1.0 pound
acquires an acceleration of 1.0 foot per second per
second.

Radiation units

1 BECQUEREL   The becquerel is an SI unit, being the activity of a
source of ionizing radiation and the activity of a
radionuclide.

1 CURIE   37,000.00000 (e) megabecquerel

The curie, named for Marie Curie (1867–1934),
Polish-born French chemist, is a unit of mass of radiation emanation. It is defined as the amount in equi-
librium with 1.0 gram of radium, as that amount
of any radioactive substance which emits the same
number of alpha rays per unit of time as 1.0 gram
of radium, and as the amount of any nuclide that
undergoes exactly $3.7 \times 10^{+10}$ radioactive disintegra-
tions per second.

1 MILLIGRAM-   The milligram-hour, a unit in which the therapeutic
HOUR   dose of radium is expressed, describes exposure to
the action of 1.0 milligram of radium for 1.0 hour.

1 REM   The rem is the relative biologic effective dose absorbed in rads multiplied by a conventionally agreed-
on value.

1 SIEVERT   100.00000 (e) rem

1 ROENTGEN   The roentgen, named for Wilhelm Konrad Roentgen
(1845–1923), German physicist, is an obsolete unit
of radiation dose, described as a unit of exposure dose
of X or gamma radiation that produces, in conjunc-
tion with its secondary electron radiation, an ioniza-
tion of 1.0 electrostatic unit of charge (approximately
$2.082 \times 10^{+9}$ electron-ion pairs) in 0.001293 gram of
air. The unit of absorbed dose is the rad, which is

| 1 *Roentgen* | the absorption of 10^{-5} joules, or 100 ergs, of radiation |
| *(continued)* | energy in 1.0 gram of material. |

More simply, the roentgen is equal to the quantity of ionizing radiation that produces 1.0 electrostatic unit of electricity in 1.0 cubic centimeter of dry air at a temperature of 0 degree centigrade and under standard atmospheric pressure.

1 GRAY   100.00000 (e) rad

The gray, named for English radiobiologist Louis H. Gray, is the SI unit of absorbed dose, equal to the amount of ionizing radiation absorbed when the energy imparted to matter is 1 joule per kilogram.

1 COULOMB PER   38,759.70000 (a) roentgen
KILOGRAM

Viscosity units

1 POISE   100.00000 (e) centipoises

The poise, the cgs (centimeter-gram-second) unit of absolute viscosity, is defined as the viscosity of a fluid that would require a shearing force of 1.0 dyne to move 1 square centimeter area of either of two parallel layers of the fluid, 1.0 centimeter apart, with a velocity of 1.0 centimeter per second relative to the other layer, the space between the layers being filled with the fluid.

1 PASCAL-   10.00000 (e) poise
SECOND

1 STOKE   100.00000 (e) centistokes

The stoke, the unit of kinematic viscosity, defines the kinematic viscosity of a fluid that has a viscosity of 1.0 poise and a density of 1.0 gram per cubic centimeter.

1 PERM   The perm represents the degree of retardation of moisture movement; the lower the value, the greater the retardation. As a unit of vapor permeance, the perm equals 57.4525 pascal/second/square meter. As a unit of vapor permeability, the perm equals 1.45929 pascal/second/meter.

Units expressing flow of water

The following units are presented with the unit allowing the least amount of water flow first, proceeding then to the last unit with the greatest flow of water.

1 PINT PER MINUTE
0.125000 (e) gallon per minute
0.071429 (r) water-inch

1 GALLON PER MINUTE
8.00000 (e) pints per minute
0.571429 (r) water-inch
0.063088 (r) liter per second

1 WATER-INCH
14.00000 (e) pints per minute
1.75000 (e) gallons per minute
0.234039 (r) cubic foot per minute
0.110404 (r) liter per second

The water-inch is a unit of hydraulics equal to the discharge from a circular, sharp-edged orifice 1.0 inch in diameter with a head of 1.0 line above the top edge.

The water-inch is also described as that quantity of water, nearly 500 cubic feet, that is discharged in 24 hours through a circular opening 1.0 inch in diameter leading from a reservoir in which the water is constantly only high enough to cover the orifice; in this case a water-inch equals 0.347222 (a) cubic foot per minute.

1 CUBIC FOOT PER MINUTE
7.47762 (r) gallons per minute
4.27280 (r) water-inches
0.833333 (r) miner's inch
0.471748 (r) liter per second

1 MINER'S INCH
1.20000 (e) cubic feet per minute
0.566324 (r) liter per second
0.020000 (e) cubic foot per second

The miner's inch equals the horizontal flow of water through a hole 1.0 inch square in a 2.0-inch plank at a depth of 4.0 inches of water.

1 LITER PER SECOND
126.80256 (r) pints per minute
15.85032 (r) gallons per minute

1 Liter per	9.05733 (r) water-inches
second	1.76577 (r) miner's inches
(continued)	0.060002 (r) cubic meter per minute

1 CUBIC METER	264.16427 (r) gallons per minute
PER MINUTE	35.31303 (r) cubic feet per minute
	16.66620 (r) liters per second
	0.588551 (r) cubic foot per second

1 CUBIC FOOT	60.00000 (e) cubic feet per minute
PER SECOND,	50.00000 (e) miner's inches
SECOND-FOOT	28.31622 (r) liters per second

The unit is usually used to describe the flow of streams.

1 CUBIC METER	15,849.85619 (r) gallons per minute
PER SECOND	2,118.78180 (r) cubic feet per minute
	999.97200 (r) liters per second
	35.31303 (r) cubic feet per second

Speed units

The following units of speed or velocity are presented with the slowest first, proceeding to those with the highest velocity.

1 FOOT PER	0.200000 (e) inch per minute
HOUR	

1 INCH PER	5.00000 (e) feet per hour
MINUTE	0.083333 (r) feet per minute

1 FOOT PER	60.00000 (e) feet per hour
MINUTE	0.304800 (r) meter per minute

1 METER PER MINUTE
196.85040 (r) feet per hour
39.37008 (r) inches per minute
3.28084 (r) feet per minute

1 KILOMETER PER HOUR
16.66667 (r) meters per minute
0.621371 (r) mile per hour
0.539587 (r) knot
0.277778 (r) meter per second

1 FOOT PER SECOND
60.00000 (e) feet per minute
0.304800 (r) meter per second

1 MILE PER HOUR
1.60935 (r) kilometers per hour
0.868383 (r) knot

1 KNOT
1.85327 (r) kilometers per hour
1.15157 (r) miles per hour
0.514806 (r) meter per second

1 METER PER SECOND
3.60000 (e) kilometers per hour
3.28084 (r) feet per second
2.23693 (r) miles per hour
1.94248 (r) knots

MACH 1
A Mach number indicates the ratio of the speed of an object to the speed of sound (about 1,088 feet per second through air at sea level at 32 degrees Fahrenheit) in the surrounding medium. Mach 1, then, means once the speed of sound; a hypersonic speed means Mach 5 or faster, or at least five times the speed of sound. The unit was named for Ernst Mach (1838–1916), Austrian physicist and philosopher.

1 GAL
0.010000 (e) meter per second per second

Miscellaneous units

1 PHONON
The phonon is the quantum of acoustic or vibrational energy, considered a discrete particle and used espe-

1 Phonon
(continued)

cially in mathematical models to calculate thermal and vibrational properties of solids.

1 VAR-HOUR

The var-hour means a reactive volt-ampere-hour, as determined by the American Standard Association in 1941. The unit expresses the integral of reactive power in vars over an interval of time in hours.

1 WATT PER STERADIAN

The watt per steradian is the solid-angle density unit of radiant power. It indicates the radiant power emitted by any source in relation to any given solid angle in a given direction.

1 MOLECULE, GRAM MOLECULE, MOLE, MOL, MOLAR, MOLAL, MOLAR WEIGHT

The molecule or a gram molecule is the smallest physical unit of an element or compound, consisting of one or more like atoms in the first case, and two or more different atoms in the second case. It is a quantity of a substance whose weight, measured in any chosen unit, is numerically equal to the molecular weight.

The mole, a unit of content, is defined as equaling 22.4 liters of gas at standard conditions of temperature and pressure, and as equaling Avogadro's number of $6.02 \times 10^{+23}$ molecules. Molecules in the air, for instance, beat against a windowpane at the rate of about 2 million billion billion per square inch per second. The average molecule moves about 1,000 miles per hour and collides with other molecules about 5 billion times per second.

The molal is described as a unit pertaining to a solution containing 1 mol of solute per liter of solution. A millimole is 1/1000 mole, and a decimolar is 1/10 molar.

1 WATER EQUIVALENT

The water equivalent is the product of the mass of a body and its specific heat, and equals numerically the mass of water equivalent in thermal capacity to the body in question, or the weight of water that would be heated to the temperature attained in a calorimeter by the heat absorbed by the instrument.

1 SIEMENS, SIEMENS MERCURY UNIT

This is considered a unit of both electrical resistance and conductance. It is defined as that resistance encountered or that conductance allowed in a column of mercury 1.0 meter high and with a cross section of

1.0 square millimeter, at a temperature of 0 degree centigrade.

The unit was named for Alfred Ernst Werner von Siemens (1816–1892), a German inventor.

Units of various capacities and volumes per various masses and weights

1 CUBIC FOOT PER (U.S.) POUND
0.074809 (r) gallon per hundredweight
0.012844 (r) cubic meter per kilogram

1 (U.S.) GALLON PER HUNDREDWEIGHT
13.36742 (r) cubic feet per pound
0.171703 (r) cubic meter per kilogram

1 CUBIC METER PER KILOGRAM
77.85181 (r) cubic feet per pound
5.82458 (r) gallons per hundredweight
0.031244 (r) bushel per short ton

1 (U.S.) BUSHEL PER SHORT TON
2,491.700000 (a) cubic feet per pound
186.40096 (r) gallons per hundredweight
32.00433 (r) cubic meters per kilogram

Units of various lengths per various capacities and volumes

1 FOOT PER (U.S.) BUSHEL
0.415283 (r) yard per cubic foot
0.010753 (r) kilometer per liter

1 YARD PER CUBIC FOOT
2.40800 (r) feet per bushel
0.025894 (r) kilometer per liter

1 KILOMETER PER LITER
92.98419 (r) feet per bushel
38.61895 (r) yards per cubic foot
0.164149 (r) mile per gallon

| 1 (STATUTE) MILE PER (U.S.) GALLON | 566.51959 (r) feet per bushel
235.26624 (r) yards per cubic foot
6.09202 (r) kilometers per liter |

Units of various masses and weights per various areas

| 1 (U.S.) POUND PER SQUARE FOOT | 0.042140 (r) kilogram per square meter |

| 1 KILOGRAM PER SQUARE METER | 23.73023 (r) pounds per square foot |

| 1 HUNDRED-WEIGHT PER SQUARE YARD | 900.00000 (e) pounds per square foot
37.92626 (r) kilograms per square meter |

| 1 SHORT TON PER ACRE | 3,671,262.06480 (a) kilograms per square meter
9,680.00000 (e) hundredweights per square yard |

Units of various capacities and volumes per various areas

| 1 (U.S.) GALLON PER SQUARE YARD | 0.111769 (r) cubic foot per square meter |

| 1 CUBIC FOOT PER SQUARE METER | 8.94702 (r) gallons per square yard |

1 (U.S.) BUSHEL PER ACRE	4,511.04166 (r) gallons per square yard 0.142608 (r) hectoliter per hectare
1 HECTOLITER PER HECTARE	31,632.32871 (r) gallons per square yard 3,535.52350 (r) cubic feet per square meter

Miscellaneous units relating to temperature

1 BTU-INCH PER SQUARE FOOT-DEGREE FAHRENHEIT	0.1442279 (r) watt per meter Kelvin
1 BTU PER HOUR SQUARE FOOT DEGREE FAHRENHEIT	5.678263 (r) watt per square meter Kelvin The SI coefficient of heat transfer.
1 BTU PER DEGREE RANKINE	1.899108 (r) kilojoule per Kelvin The SI unit of entropy.
1 BTU PER POUND DEGREE RANKINE	4.18680 (r) kilojoule per kilogram Kelvin The SI unit of specific entropy.
1 BTU PER POUND	2.32600 (a) kilojoule per kilogram The unit of specific internal energy.

PART EIGHT

Electrical units

Electrical units are presented in Part Eight:

Electrical power units

Electrical cycle units

Electrical frequency units

Electromotive force units

Electrical intensity units

Electrical resistance units

Electrical conductance units

Electrical quantity units

Electrical capacity units

Electrical inductance units

Electrical units of magnetic flux

Electrical units of magnetic flux density

Magnetomotive force units

Additional electrical units involving units of power per various areas, volt and ampere-turn per various lengths, ampere and coulomb per various areas, and ohm and mho per various volumes

The following definitions are offered for clarification and may be applicable to several sections in Part Eight.

An *electromagnetic unit* is not a unit of measure, but rather a system of units for electrical magnetism based on a system of equations in which the permeability of free space is taken as unity and by means of which the *abampere* is defined as the fundamental unit of current.

The term *per second per second* merely means per second every second, or the rate of acceleration over an indefinite period.

Electrical power units

1 ERG PER SECOND   0.000100 (e) milliwatt

1 MICROWATT   10.00000 (e) ergs per second

1 MILLIWATT   0.001000 (e) watt

1 FOOT-POUND PER MINUTE   0.022597 (r) watt
0.016667 (r) foot-pound per second

The unit is also considered to equal 1.3563 watts.

1 KILOGRAM- METER PER MINUTE   7.23671 (r) foot-pounds per minute
0.163528 (r) watt

1 BTU PER HOUR   0.293000 (a) watt

1 WATT, ABSOLUTE WATT, JOULE-SECOND, VOLT-AMPERE   44.28787 (r) foot-pounds per minute
6.11508 (r) kilogram-meters per minute
3.41297 (r) btus per hour

The watt, named for James Watt (1736–1819), Scottish inventor, is the work done at the rate of 1.0 joule per second, or the work represented by a current of 1.0 ampere under a pressure of 1.0 volt.

1 INTERNATIONAL WATT   1.00017 (r) watts

1 FOOT-POUND PER SECOND
60.00000 (e) foot-pounds per minute
1.35580 (r) watts

1 KILOGRAM-METER PER SECOND
60.00000 (e) kilogram-meters per minute
9.80480 (r) watts

1 BTU PER MINUTE
60.00000 (e) btus per hour
17.58000 (a) watts

The unit is also considered to equal 63.24 kilowatts.

1 KILOGRAM CALORIE PER MINUTE
69.71667 (r) watts

1 METRIC HORSEPOWER, FRENCH HORSEPOWER, FORCE DE CHEVAL, CHEVAL-VAPEUR
2,509.83000 (r) btus per hour
735.35992 (r) watts
75.00000 (e) kilogram-meters per second
0.986147 (r) electrical horsepower

1 BRITISH HORSEPOWER
4,559.06953 (r) kilogram-meters per minute
2,544.54000 (r) btus per hour
745.54464 (r) watts
1.01385 (r) metric horsepowers

The British horsepower is also considered to equal 4,562.43 kilogram-meters per minute and 2,546.5 btus per hour.

1 ELECTRICAL HORSEPOWER, HORSEPOWER
33,000.00000 (e) foot-pounds per minute
745.69000 (r) watts
550.00000 (e) foot-pounds per second
1.01405 (r) metric horsepower

The electrical horsepower was devised by James Watt (1736–1819), Scottish inventor of the steam engine.

A machine horsepower is considered to equal the strength of 4.4 horses, the strength of 1 horse being equivalent to that of 5 men; a draft horse can draw 1,600 avoirdupois pounds 23 statute miles per day, weight of carriage included.

Horsepower is usually determined in one of two ways. Indicated horsepower is power developed in the cylinder of an engine as calculated from (1) the average pressure of the working fluid, (2) the piston area, (3) the stroke, and (4) the number of working strokes per minute. Brake horsepower is the actual power given out by an engine calculated from (1) the force exerted on a friction brake or absorption dynamometer acting on the flywheel or brakewheel rim, (2) the effective radius of this force, and (3) the speed of the flywheel or brakewheel.

The horsepower is also an arbitrary misnamed unit of comparison for steam boilers, adopted by the Centennial Commission, and equal to the evaporation of 30 avoirdupois pounds of water per hour when the temperature of the feed water is 100 degrees Fahrenheit and the pressure of the steam is 70 pounds per square inch, as read from the gauge. It is also now defined as the equivalent of 34.5 avoirdupois pounds of steam evaporated per hour at 212 degrees Fahrenheit.

1 PONCELET

980.48000 (r) watts

100.00000 (e) kilogram-meters per second

The unit equals g watts, when g is the value of the acceleration of gravity in centimeters.

1 KILOWATT,
KILOVOLT-
AMPERE

1,000.00000 (e) watts

1.34104 (r) electrical horsepowers

0.948047 (r) btu per second

The kilowatt is also considered to equal 0.9486 btu per second.

1 BTU PER
SECOND

1,054.80000 (a) watts

777.99081 (r) foot-pounds per second

60.00000 (e) btus per minute

1 REFRIGERA- TION TON	3.51685 (r) kilowatts

1 KILOGRAM CALORIE PER SECOND	4,183.00000 (a) watts 60.00000 (e) kilogram calories per minute

Electrical cycle units

One of the highest man-made rotary speeds was achieved at the University of Virginia in 1961: 1,500,000 revolutions per second on a steel rotor with a diameter of about 1/100 inch suspended in a vacuum. The edge of the rotor traveled about 2,500 miles per hour and was subject to a pressure of about 1 billion *g*.

1 ELECTRICAL DEGREE	0.500000 (e) degree 0.017455 (r) electrical radian

1 DEGREE	2.00000 (e) electrical degrees 0.011111 (r) alternation

The term degree is also used in Linear Measure, Part One, and in Angular Measure, Part Two.

1 ELECTRICAL RADIAN	57.29578 (r) electrical degrees 28.64789 (r) degrees 0.318310 (r) alternation 0.159155 (r) electrical cycle

1 ALTERNATION, QUADRANT	180.00000 (e) electrical degrees 90.00000 (e) degrees 3.14184 (r) electrical radians 0.500000 (e) electrical cycle

The term quadrant also occurs under Angular Measure in Part Two.

1 ELECTRICAL CYCLE, CYCLE	360.00000 (e) electrical degrees 180.00000 (e) degrees 6.28319 (r) electrical radians 2.00000 (e) alternations 0.500000 (e) revolution
1 REVOLUTION	720.00000 (e) electrical degrees 360.00000 (e) degrees 12.56637 (r) electrical radians 2.00000 (e) electrical cycles
1 KILOCYCLE	1,000.00000 (e) electrical cycles
1 MEGACYCLE	1,000.00000 (e) kilocycles
1 KILO-MEGACYCLE	1,000.00000 (e) megacycles

Electrical frequency units

1 HERTZ	0.000001 (e) megahertz

The hertz, named for Heinrich Rudolf Hertz (1857–1894), German physicist, was formerly known as a cycle per second.

1 MEGAHERTZ	1,000,000.00000 (e) hertz
1 TERAHERTZ	10^{+12} (e) hertz

Electromotive force units

1 ABVOLT	10^{-8} (e) volt 0.010000 (e) microvolt
1 MICROVOLT	100.00000 (e) abvolts 0.001000 (e) millivolt

---

1 MILLIVOLT   0.001000 (e) volt

1 STATVOLT   0.003333 (r) volt

1 VOLT,   1,000.00000 (e) millivolts
ABSOLUTE VOLT,   300.00000 (e) statvolts
JOULE-COULOMB   0.999660 (r) international volt
                         10^{-9} (e) gigavolt

> The volt was named for Alessandro Volta (1745–1827), Italian physicist. According to the 1893 International Electrical Congress, the volt is that electromotive force which, steadily applied to a conductor where resistance is 1.0 ohm, produces a current of 1.0 ampere.
>   More simply: 1 volt = 1 joule/1 coulomb, and 1 volt = 10^{+7} ergs/(3 × 10^{+9})/statcoulombs. In 1948, the absolute volt, measured at 20 degrees Centigrade, of 1.018366 of the above-defined volts was again redefined to establish the international volt equaling 1.00034 of the 1948 volt, and in 1969, another change made the absolute volt equal 1.000010 of the international volt.

1 INTERNATIONAL 1.00034 (r) volts
VOLT
> The international volt is also considered to equal 1.000330 volts.

1 KILOVOLT   1,000.00000 (e) volts

1 GIGAVOLT   1,000,000.00000 (e) kilovolts

Electrical intensity units

1 STATAMPERE $\frac{1}{3}$ × 10^{-9} (e) ampere

1 MICROAMPERE   3,000.00000 (e) statamperes

1 MILLIAMPERE   1,000.00000 (e) microamperes
                         0.001000 (e) ampere

---

1 INTERNATIONAL AMPERE, INTERNATIONAL COULOMB PER SECOND

0.999835 (r) ampere

The international ampere was defined by the International Electrical Congress in 1893 as one tenth of the unit of current of the cgs (centimeter-gram-second) system of electromagnetic units or the practical equivalent of the unvarying current which, when passed through a standard solution of nitrate of silver in water, deposits silver at the rate of 0.001118 gram per second.

1 AMPERE, AMP, ABSOLUTE AMPERE, COULOMB PER SECOND

$3 \times 10^{+9}$ (e) statamperes

1,000.00000 (e) milliamperes

1.00007 (r) international amperes

0.100000 (e) abampere

The ampere, named for André Marie Ampère (1775–1836), French mathematician and physicist, is defined as the steady electrical current flowing in straight parallel wires separated by a free space distance of 1.0 meter, producing a force between the wires of 2×10^{-7} newtons per meter. It is also defined as the amount of electrical current that flows through a resistance of 1.0 ohm when a potential of 1.0 volt is applied across the resistance. A simplification of Ohm's law is:

$$\text{Current (amperes)} = \frac{\text{potential (volts)}}{\text{resistance (ohms)}}$$

or

$$1.0 \text{ ampere} = \frac{1 \text{ coulomb}}{1 \text{ second}}$$

The above definitions replace the ampere value of 0.99985 of the present absolute ampere used prior to 1948.

1 ABAMPERE, BIOT

10.00000 (e) amperes

1 KILOAMPERE

1,000.00000 (e) amperes

1 MILLIOERSTED

0.785398 (r) ampere per meter

0.001000 (e) oersted

1 AMPERE PER FOOT, AMPERE-FOOT	0.304801 (r) ampere per meter

The ampere-foot, a unit employed in calculating the fall of pressure in distributing mains, is equivalent to a current of 1.0 ampere flowing through 1.0 foot of conductor.

1 AMPERE PER METER, AMPERE-METER	3.28083 (r) amperes per foot 0.012566 (r) oersted

1 OERSTED	79.57747 (r) amperes per meter

The oersted, the electrical unit of intensity of a magnetic field, was named for Hans Christian Oersted, (1777–1851), Danish physicist. Formerly a unit of reluctance which equaled 1.0 gilbert per maxwell, the oersted now equals the intensity in a vacuum at a distance of 1.0 centimeter from a magnetic pole, or $\frac{1}{4} \times 10^{+3}$ amperes per meter.

Electrical resistance units

1 OHM-CIRCULAR MIL PER FOOT	1.662426 (r) nanoohms

1 OHM, TRUE OHM, THEORETCAL OHM, BRITISH ASSOCIATION UNIT	$3 \times 10^{+9}$ (e) statohms 0.989370 (r) legal ohm 0.987083 (r) absolute ohm 0.986600 (r) international ohm 0.100000 (e) abohm

Originally, the ohm, as established in 1893, equaled a resistance to current of a column of mercury 106.3 centimeters high and 1.0 square millimeter in cross section at 0 degrees Centigrade. This was changed in 1948, the new ohm equaling 1.00049 of the original value.

The ohm now, as used above, may be defined as the resistance of a circuit in which a potential difference of 1.0 volt produces a current of 1.0 ampere. It was named for Georg Simon Ohm (1787–1854), German electrician.

1 LEGAL OHM   1.01074 (r) ohms
0.997689 (r) absolute ohm
0.997200 (r) international ohm

The legal ohm was adopted at the Paris Congress in 1884.

1 ABSOLUTE   1.01309 (r) ohms
OHM   1.00232 (r) legal ohms
0.999510 (r) international ohm

1 INTERNATIONAL   1.01358 (r) ohms
OHM   1.00281 (r) legal ohms
1.00050 (r) absolute ohms

The international ohm, adopted in 1893, equals the resistance offered to an unvarying electrical current by a column of mercury at the temperature of melting ice, 14.4521 grams in mass, of a constant cross-sectional area, and 106.3 centimeters in length.

1 ABOHM   10.00000 (e) ohms
0.010000 (e) kilohm

1 KILOHM   100.00000 (e) abohms
0.001000 (e) megohm

1 MEGOHM   1,000.00000 (e) kilohms

1 METER-   The meter-millimeter is a unit used in expressing the
MILLIMETER   resistance of electrical conductors. However, it is best described, not as a unit of measure, but as a wire 1.0 millimeter in diameter and 1.0 meter in length.

1 TERAOHM 10^{+12} (e) ohms

1 OHM-MILE   The ohm-mile is the resistance of 1.0 ohm over a distance of 1.0 statute mile.

Electrical conductance units

1 MHO   0.100000 (e) abmho

0.000001 (e) megmho

The mho is the reciprocal of the ohm, the electrical unit of resistance. The word was coined by William Thomson, Lord Kelvin (1824–1907), British scientist.

1 ABMHO   10.00000 (e) mhos

1 MEGMHO   1,000,000.00000 (e) mhos

Electrical quantity units

1 ELECTROSTATIC UNIT, STATCOULOMB $\frac{1}{3} \times 10^{-9}$ (e) coulomb

The electrostatic unit is also any of a system of units for electricity and magnetism based on a system of equations in which the permittivity of empty space is defined as unity and by means of which a fundamental unit of charge is defined.

1 MICRO-COULOMB   0.000001 (e) coulomb

1 INTERNATIONAL COULOMB   0.999835 (r) coulomb

1 COULOMB, ABSOLUTE COULOMB, AMPERE PER SECOND

$3.998 \times 10^{+9}$ (r) electrostatic units

1.00017 (r) international coulomb

0.100000 (e) abcoulomb

0.016667 (r) ampere per minute

The coulomb was named for Charles Augustin de Coulomb (1736–1806), French physicist.

Equivalent to $2.998 \times 10^{+9} \times 10^{+10}/4.80$ electrons, the coulomb is defined as (1) the quantity of electricity conveyed in 1.0 second by the current produced by an electromotive force of 1.0 volt acting in a circuit having a resistance of 1.0 ohm, (2) the quantity on the positive plate of a conductor of 1.0-farad capacity when the electromotive force is 1.0 volt, or (3) the quantity transferred by a current of 1.0 ampere in 1.0 second.

1 ABCOULOMB   10.00000 (e) coulombs

1 AMPERE PER MINUTE
180,000,000,000.00000 (e) electrostatic units
60.00000 (e) coulombs
0.016667 (r) ampere per hour

1 AMPERE PER HOUR
$108 \times 10^{+11}$ (e) electrostatic units
3,600.00000 (e) coulombs
60.00000 (e) amperes per minute
0.037308 (r) faraday

1 FARADAY
96,494.00000 (a) coulombs
1,608.23333 (r) amperes per minute
26.80389 (r) amperes per hour

The faraday, named for Michael Faraday (1791–1867), British chemist and physicist, is the quantity of electricity capable of depositing or dissolving 1.0 gram equivalent weight of a substance in electrolysis.

Electrical capacity units

1 MICROMICRO-FARAD PICOFARAD
10^{-12} (e) farad

1 STATFARAD $\frac{1}{3} \times 10^{-9}$ (e) farad

1 MICROFARAD   0.000001 (e) farad

1 INTERNATIONAL FARAD, ABSOLUTE FARAD, VOLT-COULOMB
0.999505 (r) farad

| **1 FARAD** | $3 \times 10^{+9}$ (e) statfarads |
| | 1.00050 (r) international farads |

The farad, named for Michael Faraday (1791–1867), British chemist and physicist, is equal to the capacitance of a capacitor having a charge of 1.0 coulomb on each plate and a potential difference of 1.0 volt between the plates. More simply, 1 farad = 1 coulomb/1 volt.

1 ABFARAD   1,000,000,000.00000 (e) farads

Electrical inductance units

1 ABHENRY 10^{-9} (e) henry

0.001000 (e) microhenry

1 MICROHENRY   1,000.00000 (e) abhenries

1 MILLIHENRY   1,000.00000 (e) microhenries

1 HENRY, 10^{+9} (e) abhenries
ABSOLUTE   1,000.00000 (e) millihenries
HENRY   0.999405 (r) international henry

The henry, named for Joseph Henry (1797–1878), American physicist, was defined by the International Electrical Congress in Chicago, in 1893, as the inductance of a circuit in which an electromotive force of 1.0 volt is induced by a current varying at the rate of 1.0 ampere per second and equaling $1 \times 10^{+9}$ cgs units of inductance.

1 INTERNATIONAL 1.00050 (r) henries
HENRY

1 STATHENRY $8.9876 \times 10^{+11}$ (r) henries

Electrical units of magnetic flux

1 MAXWELL, LINE, 10^{-8} (e) weber
GAUSS
CENTIMETER The maxwell, named for James Clerk Maxwell
SQUARED (1831–1879), Scottish physicist, is the flux perpendicularly intersecting an area of 1.0 square centimeter in a region where the magnetic induction is 1.0 gauss.
  The term line also occurs under Linear Measure, Cloth Measure, and Printer's Measure in Part One.

1 MICROWEBER 0.000001 (e) weber

1 KILOLINE 10.00000 (e) microwebers

1 MILLIWEBER 0.001000 (e) weber

1 MEGALINE 10.00000 (e) milliwebers

1 WEBER, VOLT- 100,000,000.00000 (e) maxwells
SECOND
 The weber, named for Wilhelm Eduard Weber (1804–1891), German physicist, is equal to the magnetic flux that, in linking a circuit of 1.0 turn, produces in it an electromotive force of 1.0 volt as it is uniformly reduced to zero within 1.0 second.

1 MAXWELL-TURN The maxwell-turn is a unit of linkage or flux turns. It is the product of the flux turn, in maxwells, and the number of turns of conductor linked with the flux.

1 MAXWELL PER The maxwell per ampere-turn, a unit of permeance,
AMPERE-TURN is defined as the permeance of a magnetic circuit in which a magnetomotive force of 1.0 ampere-turn produces a magnetic flux of 1.0 maxwell. It equals 10^{-8} webers per ampere-turn.

Electrical units of magnetic flux density

1 ATTOTESLA 10^{-18} (e) tesla

1 GAUSS, MAXWELL PER SQUARE CENTIMETER	0.000100 (e) tesla The gauss, named for Karl Friedrich Gauss (1777–1855), German mathematician, astronomer, and physicist, equals 1.0 line of induction per square centimeter.
1 TESLA, WEBER PER SQUARE METER	10.00000 (e) kilogauss The tesla was named for Nikola Tesla (1856–1943), Austrian-born American electrical engineer.
1 GAMMA	As a unit of magnetic flux density, the gamma equals 10^{-9} tesla. As a unit of mass, the gamma equals 10^{-9} kilogram.

Magnetomotive force units

1 MICROGILBERT	0.000001 (e) gilbert
1 GILBERT	0.795775 (r) ampere-turn The gilbert, named for William Gilbert (1540–1603), English physicist and court physician, equals $10/4\pi$ ampere-turn.
1 AMPERE-TURN	1.25664 (r) gilberts 0.100000 (e) abampere-turn The ampere-turn is equal to the magnetomotive force around a path linking 1.0 turn of a conducting loop carrying a current of 1.0 ampere.
1 ABAMPERE-TURN	12.56637 (r) gilberts 10.00000 (e) ampere-turns

Units of electrical power per various areas

1 WATT PER SQUARE CENTIMETER	0.000155 (r) kilowatt per square inch

| **1 KILOWATT PER** | 6,451.62540 (r) watts per square centimeter |
| **SQUARE INCH** | 0.009311 (r) electrical horsepower per square foot |

1 ELECTRICAL	692,874.61630 (r) watts per square centimeter
HORSEPOWER	107.39536 (r) kilowatts per square inch
PER SQUARE	
FOOT	

Units involving the volt per various lengths

| **1 STATVOLT PER** | 0.033333 (r) millivolt per meter |
| **CENTIMETER** | 0.001312 (r) volt per inch |

| **1 MILLIVOLT PER** | 30.00000 (e) statvolts per centimeter |
| **METER** | 0.039370 (r) volt per inch |

| **1 VOLT PER** | 25.40005 (e) millivolts per meter |
| **INCH** | |

| **1 ABVOLT PER** | 3,000.00000 (e) statvolts per centimeter |
| **CENTIMETER** | 0.100000 (e) volt per meter |

| **1 VOLT PER** | 39.37010 (r) volts per inch |
| **METER** | 10.00000 (e) abvolts per centimeter |

Units involving the ampere-turn per various lengths

| **1 GILBERT PER** | 0.795775 (r) ampere-turn per centimeter |
| **CENTIMETER** | 0.313297 (r) ampere-turn per inch |

The unit is also considered to equal 2.021 ampere-turns per inch.

| **1 AMPERE-TURN** | 1.25664 (r) gilberts per centimeter |
| **PER CENTIMETER** | 0.393701 (r) ampere-turn per inch |

| **1 AMPERE-TURN PER INCH** | 3.19186 (r) gilberts per centimeter |
| | 2.54001 (r) ampere-turns per centimeter |

The unit is also considered to equal 0.495 gilbert per centimeter.

1 ABAMPERE-TURN PER CENTIMETER	12.56637 (r) gilberts per centimeter
	10.00000 (e) ampere-turns per centimeter
	3.93701 (r) ampere-turns per inch

Units involving the ampere per various areas

| **1 STATAMPERE PER SQUARE CENTIMETER** | $\frac{1}{3} \times 10^{-9}$ (e) ampere per square centimeter |
| | 4.649991×10^{-10} (r) amperes per square inch |

| **1 AMPERE PER SQUARE CENTIMETER** | $3 \times 10^{+9}$ (e) statamperes per square centimeter |
| | 0.15500 (r) ampere per square inch |

| **1 AMPERE PER SQUARE INCH** | $9.3548762 \times 10^{+9}$ (r) statamperes per square centimeter |
| | 6.45163 amperes per square centimeter |

The unit is also considered to equal $4.65 \times 10^{+8}$ statamperes per square centimeter.

| **1 ABAMPERE PER SQUARE CENTIMETER** | 10.00000 (e) amperes per square centimeter |
| | 1.55000 (r) amperes per square inch |

Units involving the coulomb per various areas

| **1 STATCOULOMB PER SQUARE CENTIMETER** | 2.15067×10^{-9} (a) coulombs per square inch |
| | $\frac{1}{3} \times 10^{-10}$ (e) abcoulomb per square centimeter |

1 COULOMB PER SQUARE INCH	$4.65 \times 10^{+8}$ (a) statcoulombs per square centimeter 0.154991 (r) coulomb per square meter
1 COULOMB PER SQUARE CENTIMETER	$30.0018 \times 10^{+8}$ (a) statcoulombs per square centimeter 6.45200 (r) coulombs per square inch
1 ABCOULOMB PER SQUARE CENTIMETER	10.00000 (e) coulombs per square centimeter
1 COULOMB PER SQUARE METER	1,000.00000 (e) abcoulombs per square centimeter

Units involving the ohm per various volumes

1 MICROHM PER CENTIMETER CUBE	0.061023 (r) microhm per inch cube
1 MICROHM PER INCH CUBE	16.38720 (a) microhms per centimeter cube 0.106102 (r) ohm per mil-foot
1 OHM PER MIL-FOOT	154.44820 (a) microhms per centimeter cube 9.42492 (r) microhms per inch cube
1 ABOHM PER CENTIMETER CUBE	610,232.00000 (a) microhms per inch cube 64,746.62702 (r) ohms per mil-foot

Units involving the mho per various volumes

1 MHO PER MIL-FOOT	0.000015 (r) abmho per centimeter cube

1 ABMHO PER CENTIMETER CUBE	64,732.17426 (r) mhos per mil-foot
1 MEGMHO PER CENTIMETER CUBE	100,000.00000 (e) abmhos per centimeter cube 0.061023 (r) megmho per inch cube
1 MEGMHO PER INCH CUBE	106,078,321,710.00000 (a) mhos per mil-foot 16.38720 (a) megmhos per centimeter cube

Computer Units

1 BAUD	1 bit/second Used as a measure of binary data transfer, especially with modems.
1 BIT	The smallest piece of a binary number, either 1 or a 0.
1 BYTE	8.00000 (e) bits A byte is roughly equal to a roman letter or Arabic numeral.
1 KILOBYTE	1,028.00000 (e) bytes Also called K.
1 MEGABYTE	1,000,000.00000 (e) bytes Also called a meg.
1 GIGABYTE	1,000,000,000.00000 (e) bytes

(See also MEGAHERTZ, NANOSECOND)

INDEX OF UNITS

aam (spirits), 116, 119
abampere (electrical intensity), 283
abampere/square centimeter (ampere/area), 292
abampere-turn (magnetomotive force), 290
abampere-turn/centimeter (ampere-turn/length), 292
abcoulomb (electrical quantity), 287
abcoulomb/square centimeter (coulomb/area), 293
abdat (linear), 54
abfarad (electrical capacity), 288
abhenry (electrical inductance), 288
abmho (electrical conductance), 286
abmho/centimeter cube (mho/volume), 294
abohm (electrical resistance), 285
abohm/centimeter cube (ohm/volume), 293
absolute ampere (electrical intensity), 283
absolute coulomb (electrical quantity), 286
absolute farad (electrical capacity), 287
absolute henry (electrical inductance), 288
absolute joule (energy), 262
absolute ohm (electrical resistance), 285
absolute volt (electromotive force), 282
absolute watt (electrical power), 277
abvolt (electromotive force), 281

abvolt/centimeter (volt/length), 291
acaena (linear), 38; (surface), 75
acetabulum (capacity-volume), 137
achane (capacity-volume), 167
achir (surface), 75
achtel (capacity-volume), 142, 145
acre (square), 62
acre-foot (cubic), 89
acre-inch (cubic), 89
actus (linear), 45
actus major (surface), 68
actus simplex (surface), 67
adarme (weight-mass), 201
adelaide (spirits), 114
adhaka (capacity-volume), 167
admiralty mile (linear), 15; (mariner's), 20
adoulie (capacity-volume), 167
ady (linear), 56
agate (linear), 4; (printer's), 31
agate line (printer's), 32
air mile (linear), 15; (mariner's), 20
ako (capacity-volume), 166
akov (capacity-volume), 167
album (surface), 70
ale gallon (spirits), 115
alen (linear), 9, 54, 56
alin (linear), 55
allabreve (music), 245
almenn-turma (capacity-volume), 150
almude (capacity-volume), 157
aln (linear), 51
alqueire (capacity-volume), 156
alternation (electrical cycle), 280

am (capacity-volume), 164
ampere, amp (electrical intensity),
283
ampere/foot (electrical intensity),
284
ampere/hour (electrical quantity),
287
ampere/meter (electrical intensity),
284
ampere/minute (electrical quantity),
287
ampere/second (electrical quantity),
286
ampere/square centimeter
(ampere/area), 292
ampere/square inch (ampere/area),
292
ampere-turn (magnetomotive force),
290
ampere-turn/centimeter (ampere-
turn/length), 291
ampere-turn/inch (ampere-
turn/length), 292
amphimacer (verse), 247
amphura (capacity-volume), 134,
139
amunam (capacity-volume), 167
anapest (verse), 247
angstrom, angstrom unit (linear),
2,209
angula (linear), 40
anker (spirits), 116
anomalistic month (time), 237
anomalistic year (time), 238
anoman (capacity-volume), 167
antal (spirits), 116
anukabiet (linear), 55
apatan (capacity-volume), 166
apparent day (time), 235
apt (capacity-volume), 148
Arabian foot (linear), 34
aranzada (surface), 73
ardeb (dry), 100; (capacity-volume),
148
are (surface), 213
arienzo (weight-mass), 200
arpent (square), 62
arroba (spirits), 115; (capacity-
volume), 162, 166; (weight-mass),
199, 202
arshin (linear), 48, 210
arsmin (linear), 210

artaba (capacity-volume), 131, 148
artal (weight-mass), 193
artillery mil (linear), 3
as (weight-mass), 190
asanka (count), 225
asankhyeya (count), 225
ass (weight-mass), 202
assay ton (troy), 171
assbaa (linear), 34
asta (linear), 10, 56
astral year (time), 239
astronomical unit (solar), 18
astronomical year (time), 239
atlas folo (book), 227
atmosphere (pressure), 257
atomic mass unit (avoirdupois), 173
atomic second (time), 233
attic talent (avoirdupois), 178
attometer (linear), 208
attotesla (magnetic flux density),
289
auchlet (dry), 99
aum (spirits), 118
aune (linear), 52
aurure (surface), 75
azumbre (capacity-volume), 161

bag (dry), 97; (avoirdupois), 182
baht, bat (weight-mass), 205
baker's dozen (count), 223
bale (avoirdupois), 182; (paper), 226
balita (surface), 75
balthazar (spirits), 113
bandle (linear), 56
bar (pressure), 257
barge (avoirdupois), 185
barid (linear), 35
baril (capacity-volume), 140
barile (spirits), 116; (capacity-
volume), 152
barleycorn (linear), 5, 6; (troy), 171
barn (square), 59; (surface), 212
barn gallon (spirits), 115
barony (square), 65
barrel (liquid), 93; (dry), 98;
(spirits), 118; (avoirdupois), 181;
(weight-mass), 205
barril (spirits), 116; (capacity-
volume), 153
basket (dry), 94
bath (capacity-volume), 136
batman (weight-mass), 205

baud (computer), 294
becher (capacity-volume), 141
becquerel (radiation), 267
beer gallon (spirits), 115
bel (sound), 242
bema (linear), 37
beqa (weight-mass), 170
berkovets, berkowitz (weight-mass), 200
berri (linear), 14
bes (weight-mass), 190
BeV (energy), 262
biannual (time), 239
bicentenary, bicentennial (time), 241
bicron (linear), 209
biennial (time), 239
bigha (surface), 75
billion (count), 223
bimestrial (time), 238
binal, binary (count), 222
biot (electrical intensity), 283
biquarterly (time), 238
bismerpund (weight-mass), 191
bit (computer), 294
bite (linear), 10
biweekly (time), 236
biyearly (time), 239
block (square), 64; (avoirdupois), 175
board foot (cubic), 82
board of trade commercial unit (energy), 264
bocoy (capacity-volume), 167
bodge (capacity-volume), 167
boisseau (capacity-volume), 149, 167
boll (dry), 98; (avoirdupois), 180
bolt (cloth), 27
boo (linear), 46
bota (capacity-volume), 157
botchka (capacity-volume), 160
botella (capacity-volume), 167
bottle (spirits), 112
bougle decimale (light), 253
bourgeois (printer's), 32
boutylka (spirits), 112; (capacity-volume), 158
bouw (surface), 75
box (avoirdupois), 179
bozze (capacity-volume), 152
braca (linear), 53
braccio (linear), 55
brasse (linear), 56

braza (linear), 50, 53
breve (music), 245
brevier (printer's), 32
brick (avoirdupois), 175
brilliant (printer's), 31
British association unit (electrical resistance), 284
British bottle (spirits), 117
British bushel (dry), 96
British cable length (mariner's), 20
British candle (light), 252
British cup (cooking), 108
British flask (avoirdupois), 179
British fluid ounce (apothecaries' fluid), 103
British hogshead (ale, beer), (spirits), 119
British hogshead (wine) (spirits), 120
British horsepower (electrical power), 278
British inch (linear), 7
British minim (apothecaries' fluid), 101
British peck (dry), 95
British pint (apothecaries' fluid), 105
British quarter (avoirdupois), 176
British shipping ton (cubic), 85
British standard (cubic), 88
British standard candle (light), 252
British teaspoon (cooking), 107
British thermal unit, Btu (temperature), 260
British ton (avoirdupois), 184
British yard (linear), 11
Btu/degree rankine (temperature), 275
Btu/hour (electrical power), 277
Btu/hour-square foot degree fahrenheit (temperature), 275
Btu/inch-square foot degree fahrenheit (temperature), 275
Btu/minute (temperature), 261; (electrical power), 278
Btu/pound-degree rankine (temperature), 275
Btu/pound (temperature), 275
Btu/second (electrical power), 279
Btu/square foot (temperature), 261
bu (linear), 46; (square), 72; (capacity-volume), 136, 167

bucket (dry), 95
bulk barrel (cubic), 83
bunder (surface), 75, 213
bundle (weight-mass), 226; (paper), 205
bushel (dry), 96; (cooking), 109; (avoirdupois), 178
bushel/acre (capacity/area), 274
bushel/short ton (miscellaneous), 273
butt (spirits), 121
byte (computer), 294

cab (capacity-volume), 136
caba (capacity-volume), 167
caballeria (surface), 74
caban (capacity-volume), 166
cabda (linear), 34
cable, cable length (mariner's), 20
cados (capacity-volume), 133
caesura (music), 243; (verse), 246
caffiso (capacity-volume), 167
cafiz, cahiz (capacity-volume), 131, 162, 167
caja (weight-mass), 206
cajuela (capacity-volume), 167
cal (linear), 55
calendar day (time), 235
calendar month (time), 238
calendar year (time), 238, 239
calorie (temperature), 260
canada (capacity-volume), 156
Canadian inch (linear), 7
candela/square foot (light), 253
candela/square meter (light), 253
candle, candela, candlepower (light), 252
candle-foot (light), 251
candle-hour (light), 252
candle-lumen (light), 253
candle-meter (light), 251
candle/square centimeter (light), 253
candle/square inch (light), 253
cane (linear), 41
caneca (capacity-volume), 167
canna (linear), 55, 56
cannon (printer's), 33
cannon-shot distance (linear), 16
cantaro (spirits), 115
Cape inch (linear), 7
caphite (capacity-volume), 129

capicha (capacity-volume), 167
cap octavo (book), 227
caracter (weight-mass), 201
carat, carat grain (troy), 171
carga (capacity-volume), 153, 163; (weight-mass), 204
cargo ton (cubic), 85
carload (dry), 101
cask (avoirdupois), 183
cass (capacity-volume), 144
castellano (weight-mass), 201
catty (avoirdupois), 174; (weight-mass), 205
cavan (capacity-volume), 166
cawney, cawny (surface), 75
celemin (surface), 73; (capacity-volume), 161
cental (avoirdupois), 179
centare (surface), 213
centaro (capacity-volume), 162, 166
centenary (time), 240
centigrade thermal unit (temperature), 260
centiliter (capacity-volume), 215
centilunour (time), 234
centimeter (linear), 209
centimeter cube (capacity-volume), 215
centinewton (force), 265
centistere (capacity-volume), 216
centner (avoirdupois), 179
centrad (angular), 78
centuple (count), 223
centuple calorie (temperature), 260
centuria (surface), 69
century (time), 240
cequi (weight-mass), 205
chain (surveyor's), 23
chalcon (weight-mass), 188
chalder (dry), 99
chaldron (avoirdupois), 184
chang (linear), 12
charac (linear), 11, 41
charka (capacity-volume), 158
chauldron (dry), 100
chausseemeile (linear), 17
chebel (linear), 41
chek (linear), 56
cheke (weight-mass), 205
chenica (capacity-volume), 167
Cheshire acre (square), 64
chetverik (capacity-volume), 159

chetvert (capacity-volume), 160
cheval-vapeur (electrical power),
    278
ch'ien (weight-mass), 206
ch'ih (linear), 53
chilogrammo (weight-mass), 205
chinanta (weight-mass), 205
chittack, chittak (weight-mass), 205
chkalik (capacity-volume), 158
cho (mariner's), 21; (linear) 46;
    (surface), 72
choenix (capacity-volume), 132
chopine (capacity-volume), 149
choriamb (verse), 247
choryos (linear), 54
chos (capacity-volume), 133
Christiana, Christiania (cubic), 86
chronon (time), 233
chupa (capacity-volume), 166
chupak (liquid), 91
circle (angular), 80
circular inch (square), 60
circular mil (square), 59
circular mil-foot (cubic), 82
circumference (angular), 80
cistern (avoirdupois), 183
civil year (time), 239
clima (surface), 67
climax basket (dry), 94
clove (avoirdupois), 175
codo (linear), 49
coffee spoon (cooking), 106
colluthun, colothun (capacity-
    volume), 167
columbian (printer's), 33
column inch (printer's), 33
commercial bale (paper), 227
commercial bundle (paper), 226
commercial quire (paper), 226
commercial ream (paper), 226
common day (time), 235
common lunar year (time), 238
common month (time), 238
common year (time), 239
condylos (linear), 36
congius (capacity-volume), 138
comb, coom, coomb, coombe, (dry),
    99
copa (capacity-volume), 160
cord (cubic), 87
cordel (linear), 50, 56; (surface), 75
cord foot (cubic), 83

cosmic year (time), 239
coss (linear), 55, 56
cotta, cottah (surface), 75
cotula (capacity-volume), 132
coulomb (electrical quantity), 286
coulomb/kilogram (radiation), 268
coulomb/second (electrical
    intensity), 283
coulomb/square centimeter
    (coulomb/area), 293
coulomb/square inch (coulomb/area),
    293
coulomb/square meter
    (coulomb/area), 293
couplet (verse), 248
covado (linear), 53, 56
cover (surface), 75
coyang (capacity-volume), 167
cran, crann (liquid), 93
crosa (linear), 40
crotchet (music), 244
crown octavo (book), 227
cuadra (linear), 53; (surface), 75
cuarta (linear), 49; (capacity-
    volume), 143, 167
cuarteron (capacity-volume), 153,
    160
cuartilla, cuartillo (capacity-
    volume), 152, 160, 161
cubic centimeter (capacity-volume),
    215
cubic decameter (capacity-volume),
    217
cubic decimeter (capacity-volume),
    215
cubic foot (cubic), 83
cubic foot/minute (water flow), 269
cubic foot/pound (capacity-
    volume/weight-mass), 273
cubic foot/second (water flow), 270
cubic foot/square meter (capacity-
    volume/area), 274
cubic hectometer (capacity-volume),
    217
cubic inch (cubic), 82
cubic kilometer (capacity-volume),
    217
cubic meter (capacity-volume), 216
cubic meter/kilogram (capacity-
    volume/weight-mass), 273
cubic meter/minute (water flow),
    270

cubic meter/second (water flow), 270
cubic millimeter (capacity-volume), 215
cubic statute mile (cubic), 89
cubic yard (cubic), 84
cubit (linear), 9, 40, 44
cubito (linear), 10
cubitus (linear), 44
cudava (capacity-volume), 135
cuerda, cuerdo (surface), 75
culeus (capacity-volume), 139
cumbha (capacity-volume), 136
cup (cooking), 107
curie (radiation), 267
cut (wool) (cloth), 28
cwierc (capacity-volume), 155
cyathos (capacity-volume), 132
cyathus (capacity-volume), 138
cycle/second (electrical frequency), 281

dactyl (verse), 248
dain (linear), 56
daktylos, dakylos (linear), 7, 35
danda (linear), 11, 55
dan oranja (surface), 74
darat (surface), 76
daribah (capacity-volume), 148
daric (weight-mass), 206
deal (cubic), 86
deben (weight-mass), 193
decade (time), 240
decagram (weight-mass), 219
decaliter (capacity-volume), 216
decare (surface), 213
decastere (capacity-volume), 216
decempeda (linear), 44
decennary, decennium (time), 240
deciare (surface), 213
deciatine (surface), 76
decibel (sound), 242
decigram (weight-mass), 218
deciliter (capacity-volume), 216
decillion (count), 224
decilunour (time), 234
decimal candle (light), 253
decimeter (linear), 210
decimolar (miscellanous), 272
decistere (capacity-volume), 216
dedo (linear), 49
degree (linear), 17; (angular), 78; (electrical cycle), 280

degree day (temperature), 260
dekare (surface), 76
demikilo (weight-mass), 206
demisemiquaver (music), 244
demy (book), 228
demy octavo (book), 227
den (capacity-volume), 131
denaro (weight-mass), 206
denary (count), 223
denda (linear), 40
denier (cloth), 29
depa, depoh (linear), 56
dessert spoon (apothecaries' fluid), 103
dessiatine (surface), 76
dhan (weight-mass), 205
dhanush (linear), 40
diamond (printer's), 31
diaulos (linear), 39
dichas (linear), 36
digit (linear), 6
digitus (linear), 42
dimerlie (capacity-volume), 167
dimeter (verse), 247
dinero (weight-mass), 201
diobol, diobolon (weight-mass), 188
dipodie, dipody (verse), 247
dirhem (weight-mass), 205
displacement ton (cubic), 84
distich (verse), 248
dito (linear), 210
djerib (surface), 213
djuim (linear), 7, 47
dodran (weight-mass), 190
dola (weight-mass), 199
dolichos (linear), 39
dolium (capacity-volume), 139
donum (surface), 74
double (spirits), 110; (count), 222
double centner (weight-mass), 219
double magnum (spirits), 114
douzieme (linear), 56
dozen (count), 223
dracham, drachm (avoirdupois), 173
drachma (weight-mass), 188, 205
dracma (weight-mass), 201
dracontic month (time), 236
dram (avoirdupois, 173; (apothecaries'), 186; (weight-mass), 205
dramm, dramme (weight-mass), 194, 203

dreiling (capacity-volume), 142
dreissiger (capacity-volume), 166
drona (capacity-volume), 135, 167
drop (apothecaries' fluid), 102;
    (cooking), 106
drum (liquid), 93
duad, dual (count), 222
duella (weight-mass), 190
duim, duime (linear), 7, 47
duodecillion (count), 224
duodecimo (count), 223; (book), 228
duplex (count), 222
dyne (troy), 170; (force), 265
dyne-centimeter (energy), 261
dyne-seven (force), 267
dyne/square centimeter (pressure),
    256

Edinburgh firlot (dry), 99
eighteenmo (book), 228
eimer (capacity-volume), 166, 167
el (cloth), 25; (linear), 210
electrical cycle (electrical cycle), 281
electrical degree (electrical cycle),
    280
electrical horsepower (electrical
    power), 278
electrical horsepower/square foot
    (electrical power/area), 291
electrical radian (electrical cycle),
    280
electron volt (energy), 261
electrostatic unit (electrical
    quantity), 286
Elizabethan sonnet (verse), 250
elle (cloth), 25; (linear), 48
em (printer's), 34
en (printer's), 34
endere (linear), 56
engineer's chain (surveyor's), 23
engineer's link (surveyor's), 22
engjateigur (surface), 71
English (printer's), 33
English breakfast cup (apothecaries'
    fluid), 105
English ell (cloth), 26
English imperial bushel (dry), 96
English sonnet (verse), 250
English tablespoon (apothecaries'
    fluid), 103
English teaspoon (apothecaries'
    fluid), 102

envoy (verse), 246
ennead (count), 222
ephah (capacity-volume), 137
ephemeris second (time), 233
ephemeris year (time), 239
equinoctial year (time), 239
equivalent volt (energy), 261
erg (energy), 261
erg/second (electrical power), 277
es (weight-mass), 191
escropulo (weight-mass), 198
escrupulo (weight-mass), 201
estadal (surface), 73
estadio, estadios (linear), 16, 56
estado (linear), 50
exact day (time), 235
exact year (time), 239
excelsior (printer's), 31
ezba (linear), 6

faden (linear), 48
fagot (avoirdupois), 180
fall (square), 62
famn (linear), 51
fanega (surface), 76; (liquid), 93;
    (capacity-volume), 140, 153, 162,
    166
fanegada (surface), 73
fanga (capacity-volume), 143, 157
farad (electrical capacity), 287, 288
faraday (electrical quantity), 287
fardo (weight-mass), 205
farsakh, farsakn (linear), 16, 35, 56,
    57
farsang (linear), 56, 211
fass (capacity-volume), 160
fathmur (linear), 55
fathom (mariner's), 20
favn (linear), 3, 54
feddan, feddan masri (surface), 75
femtometer (linear), 208
feralin (surface), 71
ferfathmur (surface), 71
ferfet (surface), 71
ferk (capacity-volume), 130
fermi (linear), 2, 208; (square), 59;
    (surface), 212
fermila (surface), 71
ferthumiungur (surface), 70
ferrado (surface), 76
fet (linear), 3, 5, 7, 9, 17, 55
fifth (spirits), 111

finger (cloth), 25; (spirits), 110
firkin (spirits), 116; (avoirdupois), 177
firlot (dry), 96
fiscal year (time), 239
five-cut stuff (cubic), 86
fjarding (capacity-volume), 163
fjerding (surface), 70; (capacity-volume), 146
flag (square), 61
flask (avoirdupois), 178
flemish ell (cloth), 25
fluid drachm, fluid dram, fluidram (apothecaries' fluid), 102
fluid ounce (apothecaries' fluid), 104
fluid pint (apothecaries' fluid), 105
fluid scruple (apothecaries' fluid), 102
fod (linear), 5, 7, 9, 54
fodder (avoirdupois), 183
foder (capacity-volume), 164
folio (book), 227
foot (linear), 8; (mariner's), 19; (surveyor's), 22
foot/bushel (length/capacity-volume), 273
foot-candle (light), 251
foot/hour (speed), 270
foot-lambert (light), 253
foot/minute (speed), 270
foot-pound (energy), 263
foot-poundal (energy), 262
foot-pound/minute (electrical power), 277
foot-pound/second (electrical power), 278
foot/second (speed), 271
foot-ton (energy), 264
footweight (avoirdupois), 178
force de cheval (electrical power), 278
fortin (capacity-volume), 167
fortnight (time), 236
fot (linear), 5, 9, 51
fotter (linear), 57
foute (linear), 9, 47
frail (avoirdupois), 177
frasco (capacity-volume), 140
freight ton (cubic), 85
French foot (linear), 9
French footweight (avoirdupois), 178

French geographic mile (linear), 15
French horsepower (electrical power), 278
French nautical mile (linear), 14
frequency (sound), 242
fuder (capacity-volume), 167
fun (weight-mass), 195
funt (weight-mass), 197, 199
furlong (linear), 13; (surveyor's), 23
fuss (linear), 3, 5, 7, 9, 14, 17, 52, 57
fut (linear), 9
futtermassel (capacity-volume), 141

gal (speed), 271
gallon (liquid), 92; (avoirdupois), 175; (apothecaries' fluid), 105; (cooking), 109; (spirits), 114
gallon/hundredweight (capacity-volume/weight-mass), 273
gallon/minute (water flow), 269
gallon/square yard (capacity-volume/area), 274
gamma (magnetic flux density), 290
ganta (capacity-volume), 166
gantang (liquid), 92
garava (capacity-volume), 167
gariba (capacity-volume), 131
garnetz (capacity-volume), 158
garniec (capacity-volume), 155
garrafa (capacity-volume), 143
gauss (magnetic flux density), 290
gauss centimeter squared (magnetic flux), 289
gavyuti (linear), 41
geepound (force), 267
geerah (linear), 11; (cloth), 39
geographic mile (linear), 15; (mariner's), 20
geometric pace (linear), 11
ghalva (linear), 35
giarra (weight-mass), 205
gigabyte (computer), 294
gigameter (linear), 210
gigavolt (electromotive force), 282
gilbert (magnetomotive force), 290
gilbert/centimeter (ampere-turn/length), 291
gilbert/maxwell (electrical intensity), 284
gill (liquid), 90
giornata (surface), 75
gireh (linear), 11; (cloth), 24

go (surface), 72; (capacity-volume), 151
goad (linear), 57
gomari (capacity-volume), 144
gong qing (surface), 76
googol, googolplex (count), 225
gorraf (weight-mass), 205
gradus (linear), 44
grain (troy), 171; (avoirdupois), 173; (apothecaries'), 186
gram (weight-mass), 218; (force), 265
gram calorie (temperature), 260
gram calorie/square centimeter (temperature), 261
gram-centimeter (energy), 262
gram degree (temperature), 260
gramme (linear), 5
gram-meter (energy), 262
gram molecule (miscellaneous), 272
gran (weight-mass), 206
grano (weight-mass), 200
grao (weight-mass), 198
gray (radiation), 268
Grecian cubit (linear), 37
great calorie (temperature), 260
great gross (count), 223
great primer (printer's), 33
grein (avoirdupois), 173
gross (count), 223
gross ton (avoirdupois), 184
Gunter's chain (surveyor's), 23
Gunter's link (surveyor's), 21
guz (linear), 11, 41, 55; (cloth), 24

hairbreadth, hairsbreadth (linear), 4
halbe (capacity-volume), 141
half bottle (spirits), 113
half keg (spirits), 115
half note, half step, half tone (music), 245
half yard (spirits), 112
halibiu (linear), 57
hamlah (weight-mass), 193
hand (linear), 8
handbreadth, handsbreadth (linear), 7
hank (cotton, wool) (cloth), 29
hasta (linear), 10, 40
hath (linear), 10
hatt (metric length), 210
head (avoirdupois), 175

hebdomad (count), 222; (time), 236
hectare (surface), 213
hectobar (pressure), 258
hectogram (weight-mass), 219
hectoliter (capacity-volume), 216
hectoliter/hectare (capacity-volume/area), 274
hectostere (capacity-volume), 217
heer (cloth), 29
Hefner candle (light), 252
hekat (capacity-volume), 136
hekteus (capacity-volume), 133
hemicycle (angular), 80
hemidemisemiquaver (music), 244
hemiditone (music), 244
hemiekton (capacity-volume), 133
hemina (capacity-volume), 132, 138
hemine (capacity-volume), 149
heminee (surface), 76
hemistich (verse), 248
heml (weight-mass), 194
hen (capacity-volume), 147
henry (electrical inductance), 288
heptad (count), 222; (time), 240
heptameter (verse), 248
heptastich (verse), 249
heptateuch (book), 228
heredium (surface), 69
hertz (frequency), 281
hexachord (music), 245
hexad (count), 222
hexameter (verse), 248
hexastich, hexastichon (verse), 249
hexateuch (book), 228
hide (square), 64
hin (capacity-volume), 136
hiro (linear), 46
hiyak-kin (weight-mass), 196
hiyaku-me (weight-mass), 195
hogshead (liquid), 93; (avoirdupois), 183
hold (square), 64
holzklafter (capacity-volume), 166
homer (capacity-volume), 137
homestead (square), 64
horsepower (electrical power), 278
horsepower-hour (energy), 264
horsepower-year (energy), 265
hour (angular), 79; (time), 234
hu (capacity-volume), 167
hundred (count), 223
hundredweight (avoirdupois), 179

hundredweight/square yard (weight-mass/area), 274

iamb (verse), 247
immi (capacity-volume), 165
imperial (spirits), 114
imperial gallon (liquid), 92; (apothecaries fluid), 106; (spirits), 114
imperial gill (liquid), 90; (spirits), 111
imperial inch (linear), 7
imperial octavo (book), 227
imperial pint (liquid), 91
imperial quart (liquid), 91
imperial yard (linear), 11
inch (linear), 6; (cloth) 24; (printer's), 33
inch cube (cubic), 82
inch-pound (energy), 262
inch-ton (energy), 264
infantry mil (linear), 3
international air mile (linear), 15
international ampere (electrical intensity), 283
international candle (light), 252
international carat (troy), 171
international coulomb (electrical quantity), 286
international coulomb/second (electrical intensity), 283
international farad (electrical capacity), 287
International foot candle (light), 252
international geographic mile (linear), 16
international henry (electrical inductance), 288
international inch (linear), 7
international joule (energy), 263
international nautical mile (linear), 15
international ohm (electrical resistance), 285
international volt (electromotive force), 282
international watt (electrical power), 278
Irish acre (square), 64
Irish mile (linear), 15
iron (linear), 4; (cloth), 24
Italian sonnet (verse), 250

itcze (capacity-volume), 166
itinerary pace (linear), 11

jabia (surface), 76
jaob (linear), 11, 55
jarra (capacity-volume), 152
jerib (surface), 76, 213
jeroboam (spirits), 113
jigger (spirits), 110
jitro (surface), 64
jo (linear), 46
joch (surface), 64, 76
joule (energy), 262
joule-coulomb (electromotive force), 282
joule-second (electrical power), 277
journée (surface), 76
jow (linear), 11, 55
juchart, juchert (surface), 62
juger, jugerum (surface), 69
jugum (square), 64
Julian year (time), 239
jumba (linear), 57
jumfru (capacity-volume), 163
jutro (surface), 64

kabiet (linear), 55
kalpa (time), 241
kamian (weight-mass), 198, 200
kan (capacity-volume), 153
kannor (capacity-volume), 163
kannu (capacity-volume), 166
kantang (capacity-volume), 167
kantar (weight-mass), 193, 194, 205
kapetis (capacity-volume), 167
kappe (capacity-volume), 163
kappland (surface), 76
karat (troy), 171
karch (weight-mass), 204
karsha (weight-mass), 189
kartos (capacity-volume), 144
karwar (weight-mass), 206
kassabah (linear), 54; (surface), 75
kat (weight-mass), 192
katastarsko (surface), 64
kati (weight-mass), 195
keddah (capacity-volume), 147
keel (cubic), 88; (avoirdupois), 185
keg (spirits), 116; (avoirdupois), 179
kelvin (energy), 264
ken (linear), 9, 46, 56, 210
kenning (mariner's), 21

kerat (weight-mass), 192, 205
kette (linear), 210
keup (linear), 5, 55
khar (weight-mass), 193
kharouba (weight-mass), 205
khat (linear), 9
khet (linear), 54
khous (capacity-volume), 133
khvat (linear), 9, 56
kiladja (capacity-volume), 129
kilah (capacity-volume), 147
kilan (linear), 57
kilderkin (spirits), 117
kile, kileh (weight-mass), 205
kiliare (surface), 213
kilo (weight-mass), 219
kiloampere (electrical intensity), 283
kilobar (pressure), 258
kilobyte (computer), 294
kilocalorie (temperature), 260
kilocycle (electrical cycle), 281
kilodyne (force), 265
kilogram (weight-mass), 219
kilogram calorie (temperature), 260
kilogram calorie/minute (electrical power), 278
kilogram calorie/second (temperature), 261; (electrical power), 280
kilogram force, kilopond (force), 266
kilogram-meter (energy), 263
kilogram-meter/minute (electrical power), 277
kilogram-meter/second (electrical power), 278
kilogram/square centimeter (pressure), 257
kilogram/square meter (pressure), 256; (weight-mass/area), 274
kilohm (electrical resistance), 285
kiloline (magnetic flux), 289
kiloliter (capacity-volume), 216
kilolumen (light), 253
kilomegacycle (electrical cycle), 281
kilometer (linear), 211
kilometer/hour (speed), 271
kilometer/liter (length/capacity-volume), 273
kilopascal (pressure), 257
kilostere (capacity-volume), 217
kiloton (avoirdupois), 184

kilovolt (electromotive force), 282
kilovolt-ampere (electrical power), 279
kilowatt (electrical power), 279
kilowatt-hour (energy), 264
kilowatt/square inch (electrical power/area), 291
kin (weight-mass), 174, 195
kip (avoirdupois), 183
kish (surface), 76
kist (capacity-volume), 129
klafter (linear), 52, 53
knot (linear), 15; (mariner's), 20; (speed), 271
koilon (capacity-volume), 167
koku (capacity-volume), 151
kollast (capacity-volume), 165
koltunna (capacity-volume), 164
komma-ichi-da (weight-mass), 196
kona (weight-mass), 189
kop (capacity-volume), 153
kor (capacity-volume), 137
korec (surface), 76; (capacity-volume), 160
koreci (surface), 64
kornskeppa (capacity-volume), 150
korntonde (capacity-volume), 146
korntunna (capacity-volume), 150
korrel (weight-mass), 196
korzec (capacity-volume), 155
kos, koss (linear), 16
kotyle (capacity-volume), 132
kouza (capacity-volume), 144
krina (capacity-volume), 167
kujira (linear), 46
kulmet (capacity-volume), 167
kung (linear), 210
kung ch'ih (linear), 210
kung ch'ing (surface), 213
kung fen (linear), 209
kung li (linear, 211
kung mu (surface), 213
kup (linear), 57
kvint (weight-mass), 191
kwamme, kwan (weight-mass), 196
kwarta (capacity-volume), 155
kwarterka (capacity-volume), 155
kwien (capacity-volume), 168

labor (square), 65
lachsa (weight-mass), 205
lambert (light), 253

lan (linear), 57; (surface), 76
lanaz (surface), 64
land league (linear), 16
landmil (linear), 3, 17
land mile (linear), 13
langley (energy), 261
large calorie (temperature), 260
last (dry), 101; (avoirdupois), 185
lastre (capacity-volume), 140
latro (linear), 57
lea (linear), 12, 16; (cloth), 28
league (linear), 16
leaguer (capacity-volume), 168
leap year (time), 239
legal ohm (electrical resistance), 285
legal year (time), 239
legoa (linear), 16
legua (linear), 16, 50; (surface), 75
lestrad (capacity-volume), 168
li (linear), 54
libra (weight-mass), 190, 198, 201,
    206
lieue (linear), 16, 53
light-second (solar), 18
light-year (solar), 18
ligne (linear), 5, 52
liin (linear), 7, 47
lina (linear), 5
line (linear), 5; (cloth), 24;
    (printer's), 31; (surface), 75;
    (verse), 248; (magnetic flux), 289
linea (linear), 5, 48
linha (linear), 5
linie (linear), 5, 52
liniya (linear), 5, 47
linja (linear), 5
linje (linear), 5, 50
link (surveyor's), 21
lino (surface), 75
lippy (dry), 95
lispound, lispund (weight-mass),
    181, 191, 203
listred (capacity-volume), 168
liter (capacity-volume), 216
liter/second (water flow), 269
livre (weight-mass), 206
load (cubic), 83, 88; (dry), 100;
    (avoirdupois), 183
loan (surface), 76
lofstelle (surface), 76
loket (linear), 57
lokiec (linear), 55

London (cubic), 86
long dozen (count), 223
long hundredweight (avoirdupois),
    180
long primer (printer's), 32
long ton (avoirdupois), 184
long ton/square inch (pressure), 258
lood (weight-mass), 196
loof, lof (capacity-volume), 159
lot (square), 64; (weight-mass), 199
lumen (light), 252
lumen/square centimeter (light), 251
lumen/square foot (light), 251
lumen/square meter (light), 251
lunar civil month (time), 237
lunar day (time), 235
lunar month (time), 237, 238
lunar synodic month (time), 237
lunar year (time), 238
lunation (time), 237
lune (time), 234
lunour (time), 234
lut (weight-mass), 197
lux (light), 251

maal, mal (surface), 62
maass (capacity-volume), 165
maatje (capacity-volume), 153
Mach 1 (speed), 271
machine horsepower (electrical
    power), 279
magnum (spirits), 112
mahnd (weight-mass), 206
mail (avoirdupois), 180
makuk (capacity-volume), 130
mancus (troy), 172
manpower (energy), 263
mansion (linear), 42
manzana (surface), 74, 76, 213
marco (weight-mass), 201
marco real (surface), 73
marhala (linear), 57
marie-jeanne (spirits), 114
marine league (mariner's), 21
marine ri (mariner's), 21; (linear),
    47
marine ton (cubic), 85
maris (capacity-volume), 132
Mark Twain (mariner's), 20
marok (linear), 57
mass (capacity-volume), 141
massel (capacity-volume), 166

mast (troy), 173
mattaro (capacity-volume), 152
maxwell (magnetic flux), 289
maxwell/ampere-turn (magnetic
    flux), 289
maxwell/square centimeter
    (magnetic flux density), 290
maxwell-turn (magnetic flux), 289
mean calorie (temperature), 260
mean solar day (time), 235
measure (music), 243; (verse), 247
measurement ton (cubic), 85
me cate (linear), 57
medimno (capacity-volume), 144
medimnos, medimnus (capacity-
    volume), 134
medio (capacity-volume), 161
medium octavo (book), 227
megabecquerel (radiation), 267
megabyte (computer), 294
megacycle (electrical cycle), 281
megadyne (force), 266
megadyne/square centimeter
    (pressure), 257
megahertz (frequency), 281;
    (computer), 294
megaline (magnetic flux), 289
megameter (linear), 211
megaton (avoirdupois), 184
megmho (electrical conductance),
    286
megmho/centimeter cube
    (mho/volume), 294
megmho/inch cube (mho/volume),
    294
megohm (electrical resistance), 285
meile (linear), 3, 14, 17
meio (capacity-volume), 156
mel (sound), 243
merchant's pound (troy), 173
merfold (linear), 14
merice (surface), 76
meridian (printer's), 33
meridian mile (linear), 14
meripeninkulma (linear), 14
meter (linear), 210
meter-gram (energy), 262
meter-millimeter (electrical
    resistance), 285
meter/minute (speed), 271
meter/second (speed), 271
methuselah (spirits), 113

metonic cycle (time), 238
metretes (capacity-volume), 134
metrical foot (verse), 247
metric atmosphre (pressure), 257
metric carat (troy), 171
metric centner (weight-mass), 219
metric horsepower (electrical
    power), 278
metric hundredweight (weight-
    mass), 219
metric pfund (weight-mass), 219
metric quintal (weight-mass), 219
metric slug (force), 267
metric ton (weight-mass), 219
metze (capacity-volume), 142, 166
MeV (energy), 261
mho (electrical conductance), 286
mho/mil-foot (mho/volume), 293
microampere (electrical intensity),
    282
microangstrom (linear), 2
microbar (pressure), 256
microcoulomb (electrical quantity),
    286
microfarad (electrical capacity), 287
microgram (weight-mass), 218
microhenry (electrical inductance),
    288
microhm/centimeter cube
    (ohm/volume), 293
microhm/inch cube (ohm/volume),
    293
microinch (linear), 3
microliter (capacity-volume), 215
microlux (light), 251
micrometer (linear), 209
micromicrofarad (electrical
    capacity), 287
micromicron (linear), 209
micron (linear), 3, 209
microphot (light), 251
microsecond (time), 233
microtone (music), 244
microvolt (electromotive force), 281
microwatt (electrical power), 277
microweber (magnetic flux), 289
miglio (linear), 14, 211
mignonette (printer's), 32
mijl (linear), 14, 211
mil (linear), 3, 211; (angular), 78
mila (linear), 14
mila a landi (linear), 3, 17

mile (linear), 13
mile/gallon (length/capacity-
  volume), 274
mile/hour (speed), 271
mil-foot (cubic), 82
milha (linear), 14
milla (linear), 13, 14
millare (surface), 214
mille (linear), 35
mille marin (linear), 15
millennium, milliad (time), 241
milliampere (electrical intensity),
  282
milliangstrom (linear), 2
milliarium, milliary (linear), 45
millibar (pressure), 256
millibarn (square), 59; (surface), 212
millicron (linear), 3
milliequivalent (energy), 261
millier (weight-mass), 219
milligram (weight-mass), 218
milligram-hour (radiation), 267
millihenry (electrical inductance),
  288
millilambert (light), 253
milliliter (capacity-volume), 215
millilux (light), 251
millimeter (linear), 209
millimicron (linear), 3, 209
milli-millimillion (count), 225
millimole (miscellaneous), 272
milline (printer's), 32
millioctave (sound), 242
millioersted (electrical intensity),
  283
million (count), 223
milliphot (light), 251
millisecond (time), 233
millistere (capacity-volume), 215
millivolt (electromotive force), 282
millivolt/meter (volt/length), 291
milliwatt (electrical power), 277
milliweber (magnetic flux), 289
miner's inch (water flow), 269
mingelen (capacity-volume), 154
minim (apothecaries' fluid), 102;
  (music), 245
minion (printer's), 32
minute, minute of the arc (angular),
  77; (time), 234
mira (surface), 76
miscal (weight-mass), 206, 219

mishara (surface), 76
miskal (weight-mass), 205
misura (capacity-volume), 152
mite (troy), 170
mkono (linear), 57
mna (weight-mass), 206
mo (linear), 45; (weight-mass), 195
modius (capacity-volume), 139
moio (capacity-volume), 143, 157
mol, molal, molar, molar weight,
  mole, molecule (miscellaneous),
  272
momme (weight-mass), 195
monostich (verse), 248
month (time), 238
moolum (linear), 10
moosa (weight-mass), 206
moot (linear), 8
morg, morga (surface), 76
morgen (surface), 63
mot yka (surface), 74
mou (linear), 209
moule (capacity-volume), 166
moyo (capacity-volume), 162
mu (linear), 54; (surface), 76
mud (surface), 76, 216
mug (spirits), 111
muid (capacity-volume), 165
mulvelyk (linear), 7
mushti (capacity-volume), 135
mutchkin (liquid), 90
muth (capacity-volume), 142
muth massel (capacity-volume), 142
myriagram (weight-mass), 219
myrialiter (capacity-volume), 217
myriameter (linear), 211
myriare (surface), 214

nail (cloth), 24
nanometer (linear), 209
nanosecond (time), 233
natural year (time), 239
nautical mile (linear), 15;
  (mariner's), 20
nebuchadnezzar (spirits), 114
neper (sound), 242
net ton (avoirdupois), 184
newton (force), 266
newton meter (energy), 262
newton/millimeter (energy), 262
newton/square meter (pressure),
  256

ngan (square), 62
nin (linear), 57
nisf keddah (capacity-volume), 147
niu (linear), 5, 55
niyo (weight-mass), 195
nodical month (time), 236
noggin (liquid), 90; (spirits), 111
nonillion (count), 224
nonpareil (printer's), 31
nonuple (count), 222
novemdecillion (count), 225
nylast (weight-mass), 203
nymil (linear), 211

obol, obolos, obolus (weight-mass),
    188
ochava (weight-mass), 201
ocque (weight-mass), 206
octad (count), 222
octameter (verse), 248
octant (angular), 79
octastich (verse), 248
octateuch (book), 228
octave (spirits), 117; (count), 222;
    (sound), 243; (music), 243; (verse),
    249
octavillo (capacity-volume), 161
octavo (book), 227
octennial, octennium (time), 240
octet (verse), 249
octillion (count), 224
octodecillion (count), 225
octodecimo (book), 228
octonary, octuple (count), 222
oersted (electrical intensity), 284
ogdoad (count), 222
ohm (electrical resistance), 284
ohm-circular mil/foot (electrical
    resistance), 284
ohm-mile (electrical resistance), 285
ohm/mil-foot (ohm/volume), 293
oitavo (capacity-volume), 156
oka, oke (capacity-volume), 144;
    (weight-mass), 193, 194, 204, 205
okia, okieh (weight-mass), 192
okshoofd (capacity-volume), 154
oltonde (capacity-volume), 146
oltunna (capacity-volume), 150
omer (capacity-volume), 136
ona (linear), 57
onca (weight-mass), 198
ons (weight-mass), 197

onza (weight-mass), 201
orgyia (linear), 38
orna (spirits), 116
ort (weight-mass), 191, 202
osmin, osmina (capacity-volume),
    159
ottavarima (verse), 249
ottinger (capacity-volume), 166
ottingkar (capacity-volume), 145
ounce (linear), 4; (cloth), 24;
    (spirits), 110; (troy), 172;
    (avoirdupois), 174; (apothecaries'),
    186
ounce force (force), 265
ounce-inch (energy), 262
oxhuvud (capacity-volume), 164
oxybaphon (capacity-volume), 132

paal (linear), 14
pace (linear), 11, 44; (square), 61
pack (avoirdupois), 182
packen (weight-mass), 200
paegl (capacity-volume), 145
pajak (capacity-volume), 159
pala (weight-mass), 189
palaiste (linear), 36
palame (linear), 210
palaz (linear), 56
paletz (linear), 47
palm (linear), 210
palmites (linear), 43
palmo (linear), 210
palmus (major, minor) (linear), 43
pantoum, pantun (verse), 249
para, parah, parrah (capacity-
    volume), 168
paragon (printer's), 33
parasang (linear), 42, 57
parsec (solar), 19
particate (square), 62
parto (weight-mass), 206
pascal (pressure), 256
pascal/second (viscosity), 268
paso (linear), 50
passo (linear), 53
passus (linear), 44
pe (linear), 5, 9, 53
pearl (printer's), 31
pearl grain (troy), 171
pechya (linear), 36, 37
peck (dry), 95; (cooking), 109;
    (avoirdupois), 176

peda (linear), 9
peninkulma (linear), 17
pennyweight (troy), 171
pentad (time), 240
pentameter (verse), 248
pentastich (verse), 249
pentatone (music), 245
perch (linear), 11; (surveyor's), 22;
    (square), 61; (cubic), 84
perche (linear), 12, 52
perfect ream (paper), 226
perm (viscocity), 268
pes (linear), 43
Petersburg (cubic), 86
Petrarchan sonnet (verse), 250
pfiff (capacity-volume), 141
pfund (weight-mass), 204
phonon (miscellaneous), 271
phot (light), 252
photon (light), 254
pica (printer's), 33
picofarad (electrical capacity), 287
picometer (linear), 209
picosecond (time), 233
picul (weight-mass), 196, 205
pie (linear), 7, 9, 49
piece (cloth), 27
pied (linear), 9, 52
pied anglais (linear), 9
pied de roi (linear), 9
pig (avoirdupois), 182
pin (liquid), 92
pint (liquid), 90; (dry), 94;
    (apothecaries' fluid), 107;
    (cooking), 108; (spirits), 111
pinte (capacity-volume), 149
pint/minute (water flow), 269
pipa (capacity-volume), 143
pipe (spirits), 121
plethron, plethrum (linear), 39;
    (surface), 76
pocket (avoirdupois), 181
poide de marc (weight-mass), 206
point (linear), 4; (printer's), 30;
    (angular), 78; (troy), 170
poise (viscosity), 268
poisson (capacity-volume), 148
pole (linear), 11; (surveyor's), 22
pollegada (linear), 7
polugarnetz (capacity-volume), 158
poluosmina (capacity-volume), 159
poncelet (electrical power), 279

pond (weight-mass), 197
pondus (weight-mass), 190
pony (spirits), 110
pood (avoirdupois), 177; (weight-
    mass), 200
postal mile (linear), 17
post octavo (book), 227
pot (capacity-volume), 145, 149, 165
pottar (capacity-volume), 150
pottle (liquid), 91; (dry), 95
pouc (linear), 5
pouce (linear), 7, 52
poud (weight-mass), 200
pound (troy), 172; (avoirdupois) 174;
    (apothecaries'), 187; (force), 266
poundal (force), 265
pound force (force), 266
pound/square foot (pressure), 256;
    (weight-mass/area), 274
pound/square inch (pressure), 257
pouruete (surface), 76
pous (linear), 37
prastha (capacity-volume), 135
pret (linear), 55
prime (linear), 6
printer's bundle (paper), 226
printer's quire (paper), 226
printer's ream (paper), 227
proof gallon (spirits), 114
pu (linear), 54
pulgada (linear), 7, 9, 49
puncheon (ale, beer) (spirits), 120
puncheon (wine) (spirits), 121
pund (weight-mass), 191, 202
punkt (linear), 57
punto (linear), 55; (weight-mass), 205
pygon (linear), 37

qahen (linear), 11
qasab (linear), 35
qasaba (surface), 75
quadrant (electrical cycle), 280
quadrato (surface), 75
quadrennial, quadrennium (time), 240
quadricentennial (time), 241
quadrillion (count), 224
quadruple, quadruplet, quadruplex
    (count), 222
quart (liquid), 91; (dry), 94;
    (cooking), 108; (spirits), 112;
    (time), 236
quartan (time), 236

quartarius (capacity-volume), 137
quartaut, quarte (capacity-volume), 149
quarter (cloth), 25; (angular), 78; (dry), 100; (avoirdupois), 171; (time), 238
quarter bottle (spirits), 113
quarter hour (time), 234
quarter keg (spirits), 115
quartern (liquid), 90; (dry), 95; (avoirdupois), 175
quarternary (count), 222
quarternion (count), 222
quarteron (surface), 76
quarter point (angular), 78
quarter sack (avoirdupois), 181
quarter section (square), 64
quarter tone (music), 244
quarter yard (spirits), 111
quarto (capacity-volume), 143, 156; (book), 227
quatrain (verse), 249
quattuordecillion (count), 225
quaver (music), 244
Quebec, Quebec deal, Quebec standard deal (cubic), 86
quentchen (weight-mass), 204
quilate (weight-mass), 200, 204, 205
quindecinnial, quindecennium (time), 240
quindecillion (count), 225
quinon (surface), 76
quinquennial, quinquennium (time), 240
quint (weight-mass), 191
quinta (avoirdupois), 179
quintal, quintal metrique (avoirdupois), 179; (weight-mass), 219
quintant (angular), 80
quintet (count), 222
quintillion (count), 224
quintin (weight-mass), 191
quinto-quadagintillion (count), 225
quintuple, quintuplet (count), 222
quire (paper), 226
quo (surface), 76

rad (radiation), 267
radian (angular), 79
rai (surface), 76
ralica, ralo (surface), 74

Ramden's chain (surveyor's), 23
Ramden's link (surveyor's), 22
ratel (weight-mass), 206
rati, ratti (weight-mass), 205
rational calorie (temperature), 260
ream (paper), 226
reed (linear), 11
ref (linear), 51
refrigeration ton (electrical power), 280
register ton (cubic), 86
regular sonnet (verse), 250
rehoboam (spirits), 113
rem (radiation), 267
remen (linear), 57
rest (music), 243
retti (weight-mass), 189
revolution (angular), 80; (electrical cycle), 281
Rhine foot (linear), 9
ri (mariner's), 21; (linear), 47
rif (linear), 56
rin (linear), 46; (weight-mass), 195
rob, roub, roubouh (capacity-volume), 147
robhah (capacity-volume), 147
rod (linear), 11; (surveyor's), 22
rode (linear), 54
rod weight (avoirdupois), 185
roede (linear), 210
roeneng (linear), 56
roentgen (radiation), 267
roll (square), 61
Roman foot (linear), 9, 43
Roman mile (linear), 45
Roman pace (linear), 11, 44
rondeau, rondel, rondelet (verse), 249
rood (square), 62
room (avoirdupois), 185
ropani (surface), 77
rope (linear), 12
roquille (capacity-volume), 148
rotl (weight-mass), 205
round stave basket (dry), 94
roupi (linear), 8
royal octavo (book), 227
ruby (linear), 5; (printer's), 31
rudlet, rundlet, runlet (spirits), 117
rute, ruthe (linear), 57

saa (capacity-volume), 130, 168
sack (dry) 98; (avoirdupois), 181

saco (weight-mass), 204
sagen, sagene (linear), 9, 48
sah (linear), 57
sahme (surface), 75
salm, salma (weight-mass), 206
salmanezer (spirits), 113
saltus (surface), 69
sao (surface), 77
sarpler (avoirdupois), 184
satlijk (weight-mass), 204
saum (weight-mass), 204
sazen (linear), 55
schepel (capacity-volume), 154, 216
schun (linear), 9, 52
score (avoirdupois), 176; (count), 223
Scottish acre (square), 62
Scottish ell (cloth), 26
Scottish mile (linear), 14
scruple, scrupple (apothecaries'), 186
scrupulus, scripulum (surface), 67; (weight-mass), 190
se (surface), 72, 77
seam (dry), 100; (avoirdupois), 180
sea mile (linear), 15; (mariner's), 20
second (angular), 77; (time), 233; (music), 243
second-foot (water flow), 270
section (square), 65
seidel (capacity-volume), 141
selamin (capacity-volume), 156
semibreve (music), 245
semicircle (angular), 80
semiditone (music), 244
semiquaver (music), 244
semitone (music), 245
semiweekly (time), 236
semiyearly (time), 239
sen (linear), 56
senarius (verse), 248
senary (count), 222
septave (music), 245
septcentenary (time), 241
septenarius (verse), 248
septenary (time), 240
septendecillion (count), 225
septendecimal (count), 223
septennate, septennial, septennium (time), 240
septet, septuple (count), 222
septillion (count), 224
septimole (music), 245

septuagenary (time), 240
ser (weight-mass), 205
sesma (linear), 49
sestet (verse), 249
sestina (verse), 246
setier (capacity-volume), 165
sexagenary (count) 223
sexcentenary (time), 241
sexdecillion (count), 225
sexennial (time), 240
sextant (angular), 80
sextarius (capacity-volume), 138
sextillion (count), 224
sextodecimo (book), 228
sextula (weight-mass), 190
sextuple (music), 244
shaftment, shathmont (linear), 8
Shakespearean sonnet (verse), 250
shaku (linear), 9, 46; (surface), 72; (capacity-volume), 151
shed (square), 59
shekel (weight-mass), 206
sheng (capacity-volume), 168
shepel (capacity-volume), 165
shi (weight-mass), 195
shipload (cubic), 89
shipping ton (cubic), 85
ship pound (avoirdupois), 181
sho (capacity-volume), 151
short ream (paper), 226
short ton (avoirdupois), 184
short ton/acre (mass-weight/area), 274
shot (spirits), 110
sicilicum (weight-mass), 190
sidereal day (time), 235
sidereal hour (time), 234
sidereal month (time), 237
sidereal year (time), 239
Siegbohn unit (linear), 2
siemens, siemens mercury unit (miscellaneous), 272
sievert (radiation), 267
sign (angular), 79
silversmith's point (linear), 3
sinzer (linear), 25
sitio (surface), 77
sixmo (book), 227
sixteenmo (book), 228
size (linear), 5
sjomil (linear), 14; (mariner's), 21
skaalpund (weight-mass), 206

skalpund (weight-mass), 202
skein (cloth), 28
skeppund (weight-mass), 203
skew (count), 225
skibslast (weight-mass), 192
skieppe (capacity-volume), 168
skippund (weight-mass), 192
skrupul (weight-mass), 197
slug (avoirdupois), 177; (force), 267
small calorie (temperature), 260
small pica (printer's), 32
sok, metric (linear), 56
solar month (time), 238
solar year (time), 239
solidus (weight-mass), 190
sonnet (verse), 249, 250
span (cloth), 25
spann (capacity-volume), 164
Spenserian sonnet (verse), 250
spindle (cloth), 30
spithame (linear), 36
splint basket (dry), 94
split (spirits), 111
split deal (cubic), 86
spondee (verse), 247, 248
square (square), 61, 64
square alen (surface), 70
square centimeter (surface), 213
square chain (square), 62
square decameter (surface), 213
square decimeter (surface), 213
square fathom (square), 61
square femtometer (surface), 212
square foot (square), 60
square furlong (square), 64
square inch (square), 60
square kafiz (surface), 213
square khvat (surface), 64
square kilometer (surface), 214
square league (square), 65
square line (square), 60
square link (square), 60
square meter (surface), 213
square mil (square), 59
square millimeter (surface), 212
square pede (surface), 67
square perch, square pole, square
   rod (square), 61
square ruten (surface), 63
square Scottish ell (square), 60
square statute mile (square), 65
square vara (surface), 73, 75

square yard (square), 60
staab (linear), 52
stab (linear), 210
stack (cubic), 87
stadian, stadium (linear), 12, 41, 45
stadion (linear), 39, 211
stand (avoirdupois), 181
standard (cubic), 83
standard atmosphere (pressure),
   258
standard candle (light), 252
standard deal (cubic), 86
standard hundred (cubic), 86
standard inch (linear), 7
stang (linear), 9, 51
stanza (verse), 249
statampere (electrical intensity),
   282
statampere/square centimeter
   (ampere/area), 292
statcoulomb (electrical quantity),
   286
statcoulomb/square centimeter
   (coulomb/area), 292
stater (weight-mass), 194
statfarad (electrical capacity), 287
stathenry (electrical inductance),
   288
stathmos (linear), 40
statute mile (linear), 13;
   (surveyor's), 23
statvolt (electromotive force), 282
statvolt/centimeter (volt/length),
   291
stave (verse), 246
staych (surface), 76
steam calorie (temperature), 260
stein (spirits), 111
sten (weight-mass), 203
step (linear), 11
steradian (angular), 80
stere (capacity-volume), 216
stero (capacity-volume), 168
stich (verse), 248
stick (printer's), 34
stimpart (dry), 95
stoke (viscosity), 268
stone (avoirdupois), 175, 176
stoop (spirits), 115
stop (capacity-volume), 163
stopa (linear), 5, 9, 55
stoup (liquid), 91

St. Petersburg standard deal
(cubic), 86
streep (linear), 209
stremma (surface), 213
strich (linear), 209
strike (dry), 97
strophe (verse), 249
struck bushel (dry), 97
stunde (linear), 52
suerte (surface), 77
suld (linear), 48
sulung (square), 64
sun (linear), 46
survey foot (linear), 9; (surveyor's),
22
surveyor's chain (surveyor's), 22
surveyor's link (surveyor's), 21
Swedish mile (linear), 3, 52
synodic month (time), 237

tablespoon (cooking), 103, 107
tagwerk, tagwert (square), 63
takar (capacity-volume), 168
talanton (weight-mass), 194
talent (avoirdupois), 178
tambor (avoirdupois), 177
tan (surface), 72
tank (liquid), 94
tankard (spirits), 111
tappit-hen (spirits), 114
tarea (surface), 77
tarri (capacity-volume), 168
tavola (surface), 75
tchast (capacity-volume), 158
teacup (apothecaries' fluid), 104
teaspoon (apothecares' fluid), 102;
(cooking), 106
teman (capacity-volume), 152, 168
tenthmeter (linear), 209
terahertz (frequency), 281
terameter (linear), 210
teraohm (electrical resistance), 285
tercentenary (time), 241
tercia (linear), 9
tercio (weight-mass), 206
termino (weight-mass), 205
ternary (count), 222
terza rima (verse), 246
tesla (magnetic flux density), 290
tetrachord (music), 245
tetrad (count), 222
tetradrachma (weight-mass), 188

tetragram (verse), 247
tetrameter (verse), 248
tetrastich (verse), 249
thangsat (capacity-volume), 168
theb (linear), 54
theoretical ohm (electrical
resistance), 284
therm (energy), 265
thermochemical calorie
(temperature), 260
thill (cubic), 86
thirtytwomo (book), 228
thou (linear), 57
thousand (count), 223
thread (cloth), 27
thumlunger (linear), 7
tical (weight-mass), 205, 206
tierce (spirits), 119; (avoirdupois),
182
timber foot (cubic), 83
timber load (cubic), 85
timber ton (cubic), 85
to (capacity-volume), , 151
tod (avoirdupois), 176
toembak (capacity-volume), 168
toise (linear), 16
tola (weight-mass), 205
toll (linear), 7, 47
toman, tomand (weight-mass), 206
tomin (weight-mass), 201
tomme (linear), 7
ton, *see* assay ton, British shipping
ton, British ton, cargo ton,
displacement ton, foot-ton,
freight ton, gross ton, long ton,
marine ton, metric ton, net ton,
shipping ton, short ton, timber
ton, ton-kilometer, ton-mile,
water ton
tonde (capacity-volume), 145, 146
tondeland (surface), 70
tonel (capacity-volume), 143
tonelada (capacity-volume), 140,
157; (weight-mass), 206
ton-kilometer (energy), 264
ton-mile (energy), 265
tonne (surface), 77; (avoirdupois),
184
tonneau (weight-mass), 219
tonneau de jauge (cubic), 87
tonneau de mer (cubic), 87
tonnstelle (surface), 77

tonos (weight-mass), 194
toop (capacity-volume), 158
topo (surface), 77
torr (pressure), 257
totchka (linear), 47
toumnah (capacity-volume), 146
tovar (weight-mass), 204
township (square), 65
transmission unit (sound), 242
tredecillion (count), 225
triad (count), 222; (music), 245
tricentennial (time), 241
triennial, triennium (time), 240
trillion (count), 223
trimester (time), 238
trimeter (verse), 248
triolet (verse), 249
triple (spirits), 111; (count), 222
triplet (music), 245; (verse), 249
triplex (count), 222
tripody (verse), 248
triseme (verse), 247
tristich (verse), 249
tritone (music), 245
trochee (verse), 247, 248
tropical month (time), 236
tropical year (time), 239
true ohm (electrical resistance),
   284
true solar day (time), 235
trug (capacity-volume), 168
truss (straw, hay) (avoirdupois),
   177, 178
tsubo (surface), 72
ts'un (linear), 53
tu (linear), 54
tub (avoirdupois), 179
tum (linear), 51
tun (surface), 77; (spirits), 122
tundagslatta (surface), 71
tunland, tunnland (surface), 77
tunna (capacity-volume), 164, 166
tunna smjors (weight-mass), 206
tuuma (linear), 7
twelvemo (count), 223; (book), 228
twofold (count), 222
typp (cloth), 29

uckia (weight-mass), 206
uncia (linear), 42; (surface), 68;
   (weight-mass), 190
uncya (weight-mass), 197

undecillion (count), 224
urn, urna (capacity-volume), 139

vara (linear), 5, 9, 10, 16, 49, 53
var-hour (miscellaneous), 272
vat (capacity-volume), 216
vedro (capacity-volume), 159
verchoc, verchok (linear), 47
verse (verse), 248
verst, versta, verste (linear), 48
versus (surface), 68
vicennial (time), 240
vierkante roede (surface), 213
viertel (capacity-volume), 142, 145,
   165
vigintillion (count), 225
vingerhoed (capacity-volume), 153
violle, violle's standard (light), 252
vitasti (linear), 40
vloka, (surface), 77
voet (linear), 9
vog (weight-mass), 192
volt (electromotive force), 282
volt-ampere (electrical power), 277
volt-coulomb (electrical capacity), 287
volt/inch (volt/length), 291
volt/meter (volt/length), 291
volt-second (magnetic flux), 289

wa, wah (linear), 55, 56
wagon (weight-mass), 204
war, wari (linear), 11
water equivalent (miscellaneous),
   272
water-inch (water flow), 269
water ton (liquid), 93
watt (electrical power), 277
watt-hour (energy), 264
watt-second (energy), 262
watt/square centimeter (electrical
   power/area), 290
watt steradian (miscellaneous), 272
weber (magnetic flux), 289
weber/square meter (magnetic flux
   density), 290
week (time), 236
wey (avoirdupois), 181
whole deal (cubic), 86
whole note (music), 245
wichje, wichtje (weight-mass), 196,
   219
Winchester bushel (dry), 97

windle (dry), 98
wine gallon (spirits), 114
wisse (capacity-volume), 154, 216
wloka (surface), 77
woibe (capacity-volume), 130
wrap (linear), 12

xestes (capacity-volume), 132
X-unit (linear), 2, 208
xylon (linear), 38

yard (linear), 10; (mariner's), 19;
(surveyor's), 22; (cloth), 26;
(spirits), 113
yard/cubic foot (length/capacity-
volume), 273

yava (weight-mass), 189
year (time), 239
yin (linear), 54, 210
yojan, yojana (linear), 55
yoke (square), 64
yote, yut (linear), 56
yugada (surface), 73
yusdrum (weight-mass), 205

zak (capacity-volume), 154, 216
zira, zirai (linear), 210
zoll (linear), 7, 52
zollpfund (weight-mass), 219
zolotnik (weight-mass), 199

GENERAL INDEX

Abyssinia: cloth measue, 25; linear
measure, 56
acoustic energy unit, 270–271
Aden: linear measure, 11
air: weight of, 258
ale: spirits measure, 111, 113, 115,
116, 117, 118, 119, 120, 121, 122
Algeria: capacity and volume units,
168; weight and mass units, 206
American Standard Asociation, 272
Ampère, André Marie, 283
Anglo-Saxons: linear measure, 11
Angola: avoirdupois weight, 179
Angström, A. J., 2
angular (circular) measure: in
Babylonia, 77; units of 77–80
Annam (Vietnam area): surface
units, 76, 77
apothecaries' fluid measure units
101–106; comparison with

cooking, cubic, dry, liquid, and
spirits measures (tables),
123–126; (size), 126–129 (intermeasure); metric units compared
with, 217–218. *See also* capacity
and volume units
apothecaries' weight: comparison
with avoirdupois and troy units,
187–188; metric mass units
compared with, 219–220; units of,
186–187
Arabia: capacity and volume units,
129–131, 168; linear measure,
9, 13, 16, 34–35, 57; surface
units, 75; weight and mass units,
206
areas: ampere (electrical unit) per
areas, 292; capacity and volume
units per areas, 274–275; coulomb
(electrical unit) per areas, 292-

293; electrical power units per areas, 290–291; mass and weight units per areas, 274

Argentina: capacity and volume units, 140; linear measure, 5, 9, 10, 13, 14, 16, 53; metric surface units, 213; surface units, 75; weight and mass units, 179, 198, 200, 202, 206

Assyria: capacity and volume units, 130, 131; linear measure, 9, 10, 35, 42

astronomy. *See* solar measure; time measure

Attica (ancient Greece): avoirdupois weight, 178

Austria: capacity and volume units, 141–142; linear measure, 5, 9, 14, 17, 57; surface units, 64, 76; weight and mass units, 180, 204, 219

avoirdupois weight: comparison with apothecaries' and troy units, 187–188; defined, 173; international pound established, 174; metric mass units compared with, 219–220; units of, 173–185

Babylonia: avoirdupois weight, 178; earliest exact measure of capacity used in, 82; planimetry, 77; sexagesimal system, xviii, xvi

Barbados: spirits measure, 114; United States customary system, xiv

Bavaria: capacity and volume units, 166; linear measure, 5, 9, 57; square measure, 63; weight and mass units, 204, 206

beef: avoirdupois weight, 181

beer: spirits measure, 111, 113, 115, 116, 117, 118, 119, 120, 121, 122

Belgium: capacity and volume units, 167; cubic measure, 87; dry measure, 101; linear measure, 9, 12, 52

Bell, Alexander Graham, 242

Bengal: surface units, 75

bible: biblical cubit, 10; book size units, 288; cloth measure, 26; Gutenberg's, 30. *See also* Israel (ancient)

binary system, xviii

Bohemia: linear measure, 9; surface units, 76

Bolivia: weight and mass units, 177, 199, 202

Bombay: capacity and volume units, 167, 168

book size: in Great Britain, 228; Old Testament, 228; in United States, 228; units of, 227–228

Borneo, North. *See* Sabah

brandy: spirits measure, 110, 113

Brazil: avoirdupois weight, 179, 182; capacity and volume units, 143; linear meaure, 7, 9, 10, 14, 16, 53; metric length units, 210; surface units, 75; weight and mass units, 179, 182, 198, 199, 200, 202, 206

bread: avoirdupois weight, 175

brick: cubic measure, 84

British imperial system of measures, 11; fundamental quantitative units, xv; vs. United States customary system, xv

Brunswick (Germany): avoirdupois weight, 179–180

Buddhist count units, 225

Bulgaria: capacity and volume units, 167; surface units, 76; weight and mass units, 194, 204

butter: avoirdupois weight, 178; cooking measure, 107; spirits measure, 116

buttons: cloth measure, 24

Byelorussian SSR: linear measure, 7, 47

Caesar, Julius, 230

Calcutta: linear measure, 8, 11, 55

calendars: Egyptian, 230; Gregorian, 230, 231; international fixed, 231; Jewish, 230; Julian, 230–231; lunar, 230; lunisolar, 230; Mayan, 230; Muhammadan, 230; perpetual, 231; solar, 230; world, 231

California: linear measure, 10

Cambodia: capacity and volume units, 167

Canada: linear measure, 7, 9. *See also* Quebec

Canary Islands: surface units, 73

capacity: comparison of

apothecaries' fluid, cooking, cubic, dry, liquid, and spirits measures (tables), 123–126 (size), 126–129 (intermeasure); defined, xvii, 82; earliest exact measure of, 82; standards, xv; volume and, 82. *See also* capacity and volume units

capacity and volume units, 81–168; in Algeria, 168; in Arabia, 129–131, 168; per areas, 274–275; in Argentina, 140; in Assyria, 130, 131; in Austria, 141–142; in Barbados, 114; in Bavaria, 166; in Belgium, 87, 101, 167; in Bombay, 167, 168; in Brazil, 143; in Bulgaria, 167; in Cambodia, 167; in Castile, 163; in Ceylon, 167, 168; in Chaldea, 130, 131; in Chile, 167; in China, 167, 168; in Colombia, 161; in Costa Rica, 167; in Cuba, 162, 167; in Cyprus, 144; in Denmark, 145–146; in East Germany, 167; in East Indies, 167, 168; in Ecuador, 167; in Egypt, 96, 100, 146–148; in El Salvador, 167; in Estonia, 158; in Finland, 166; in France, 148–149; in Gaul, 93; in Germany, 119; in Great Britain, 85, 87, 89–90, 91, 93, 97, 98, 99, 101, 113, 114, 116, 119, 121, 167, 168; in Greece, 116, 121, 131–135, 167; in Guatemala, 162; in Haiti, 114; in Hungary, 116, 166; in Iceland, 150; in India, 135–136, 167; in Iran, 167; in Israel (ancient), 113, 136–137; in Italy, 168; in Japan, 151; in Latvia, 167; length units per capacity and volume units, 273–274; in Libya, 151–152; in Lisbon (Portugal), 156; in Luxembourg, 167; in Madras, 168; in Malaysia, 91; in Malta, 167; metric, 214–217; metric compared with customary units, 217–218; in Mexico, 152–153; in Netherlands, 101, 115, 116, 119, 153–154, 216; in Nicaragua, 167; in Norway, 145, 146, 168; in Panama, 161; in Paraguay, 93, 167; in Persia, 130, 131, 167; in Philippines, 166; in Poland, 155; in Portugal, 155–157; in Prussia, 101; in Rome (ancient),

116, 134, 137–140; in Rumania, 167; in Russia, 112, 157–160; in Sabah, 92; in Scotland, 96, 99; in Somaliland, 167; in South Africa, 165, 168; in Spain, 115, 160–163; in Sweden, 163–165; in Switzerland, 165–166; in Syria, 167; in Thailand, 168; in Trieste, 116; in Turkey, 167; in United States, 86, 101; in Uruguay, 166; in Venezuela, 162, 166–167; in Wales, 168; per weight and mass units, 273; in West Germany, 167; in Yugoslavia, 167

Cape Town University, South Africa, 225

Celsius, Anders, 259

Celsius temperature scale, xvii, 259

Celtic usage: dry measure, 98

cement: weight and mass units, 205; avoirdupois weight, 181, 182

Centennial Commission (England), 279

centigrade. *See* Celsius

centimeter-gram-second subsystem (cgs), 252

Ceylon: capacity and volume units, 167, 168

cgs. *See* centimeter-gram-second subsystem

Chaldea: capacity and volume units, 130, 131; linear measure, 35, 42

champagne: spirits measure, 111, 113

cheese: avoirdupois weight, 175

Chile: avoirdupois weight, 179; capacity and volume units, 167; linear measure, 5, 10, 16, 53; surface units, 75; weight and mass units, 179, 199, 200, 202

China: avoirdupois weight, 174; capacity and volume units, 167, 168; decimal system, xviii; linear measure, 53–54; metric length, 210, 211; metric surface units, 213; printer's measure, 30; surface units, 76; weight and mass units, 174, 206

circular measure. *See* angular measure

cloth measure units, 23–30; in
Abyssinia, 25; in Estonia, 25; in
France, 29; in India, 24; in
Surinam, 25; in Switzerland, 25.
See also bible
Coal (hard coal; soft coal): avoirdupois weight, 181, 184, 185; cubic
measure, 89; dry measure, 100,
101
cocoa: avoirdupois weight, 183
coffee: avoirdupois weight, 182
coins, 171; troy weight, 170
coke: avoirdupois weight, 184; cubic
measure, 89; dry measure, 100
College of Physicians (England), 186
Colombia: avoirdupois weight, 179,
182; capacity and volume units,
161; linear measure, 10; weight
and mass units, 179, 182, 198,
200, 202, 204
Comte, Auguste, 231
cooking measure units, 106–109;
comparison with apothecaries'
fluid, cubic, dry, liquid, spirits
measures (tables), 123–126 (size),
126–129 (intermeasure). *See also*
capacity and volume units
Copenhagen (Denmark): Olsen
clock, 233
Costa Rica: avoirdupois weight, 182;
capacity and volume units, 167;
surface units, 74, 76; weight and
mass units, 182, 206
cotton: avoirdupois weight, 175, 181,
182; cloth measure, 27, 28, 29, 30;
de Coulomb, Charles Augustin,
286
count units, 222–225; in France,
224; in Germany, 224; in Great
Britain, 224; in United States,
223, 224
cranberries: dry measure, 98
Cromwell, Oliver, 223
Cuba: capacity and volume units,
162, 167; linear measure, 10, 16,
56; surface units, 74, 75, 77;
weight and mass units, 199, 202
cubic measure units, 82–89;
comparison with apothecaries'
fluid, cooking, dry, liquid, and
spirits measures (tables), 123–126
(size), 126–129 (intermeasure);

metric units compared with,
217–218. *See also* capacity and
volume units
Cumbrae Supply Company, 110
Curie, Marie, 267
Cyprus: capacity and volume units,
144; linear measure, 8; surface
units, 74; weight and mass units,
193, 194, 206
Czechoslovakia: linear measure, 57;
surface units, 64, 76

dead load defined, xvii
decimal system, xvii, xviii
Delambre, J. B. J., 210
Denmark: avoirdupois weight, 179,
180; capacity and volume units,
145–146; linear measure, 3, 5, 7,
9, 17, 54; mariner's measure, 20;
surface units, 69–70; weight and
mass units, 179, 180, 191–192,
202, 203. *See also* Copenhagen
derived unit defined, xv
diamonds: troy weight, 170
Dominican Republic: linear
measure, 57; surface units, 77
dry measure units, 94–101;
comparison with apothecaries'
fluid, cooking, cubic, liquid, and
spirits measures (tables), 123–126
(size), 126–129 (intermeasure);
metric units compared with,
217–218; origin of system, 90. *See
also* capacity and volume units;
cooking measure units
Dunkirk (France), 210
duodecimal system, xviii
Dutch East Indies: linear measure,
56, 57; surface units, 75

Earth (planet): time measure and,
228–229
earth (soil): cubic measure, 83, 88.
See also land
East Africa: linear measure, 57
East Germany: capacity and volume
units, 167
East Indies: capacity and volume
units, 167, 168
Ecuador: capacity and volume units,
167; linear measure, 53; weight
and mass units, 199, 202

Edgar, King (England): cloth measure, 26
Edward I (England): avoirdupois weight, 173
Edward II (England): linear measure, 6
Edwards, Willard E., 231
Egypt: avoirdupois weight, 182; capacity and volume units, 96, 100, 146–148; decimal system, xviii; earliest known weight standard, 170; linear measure, 10, 54, 57; surface units, 75; weight and mass units, 182, 192–194
El Salvador: avoirdupois weight, 179; capacity and volume units, 167; linear measure, 9, 10; surface units, 74; weight and mass units, 179, 206
electrical units, 277–294; ampere per areas, 292; ampere-turn per length units, 291–292; coulomb per areas, 292–293; electrical capacity units, 287–288; electrical conductance units, 272–273, 286; electrical cycle units, 280–281; electrical frequency units, 281; electrical inductance units, 288; electrical intensity units, 282–284; elctrical power units, 277–280; electrical power units per areas, 290–291; electrical quantity units, 286–287; electrical resistance units, 272–273, 284–285; electromagnetic unit system defined, 277; electromotive force units, 281–282; magnetic flux, 289; magnetic flux density, 289–290; magnetomotive force, 290; mho per volume units, 293–294; ohm per volume units, 293; *per second per second* defined, 277; rotary speed record, 280; volt per length units, 291
electricity: linear measure, 2
Elizabeth I (England): avoirdupois weight, 180; linear measure, 13
energy units, 261–265. *See also* electrical units
England. *See* Great Britain
Estonia: capacity and volume units, 158; cloth measure, 25; linear measure, 7, 47, 48; surface units, 76, 77
Ethiopia. *See* Abyssinia

Fahrenheit, Gabriel Daniel, 259
Fahrenheit temperature scale, xvii, 259
Fair Packaging and Labeling Act, U.S., xiv
Faraday, Michael, 287, 288
Fermi, Enrico, 2, 208, 212
Finland: capacity and volume units, 166; linear measure, 7, 14, 17, 48; mariner's measure, 21; surface units, 77; weight and mass units, 203
firearms: angular measure, 78; linear measure, 3, 4
fish: avoirdupois weight, 181
flax: avoirdupois weight, 182, 184
flooring: square measure, 61
flour: avoirdupois weight, 176, 181; cooking mearue, 107
fluid measure. *See* apothecaries' fluid measure units; liquid measure units; spirits measure units
force units, 265–267; in Great Britain, 266, 267
France: avoirdupois weight, 184; capacity and volume units, 148–149; cloth measure, 29; count units, 224; international system of measure, xiii; linear measure, 4, 5, 7, 9, 14, 16, 53, 56; mariner's measure, 20; surface units, 63, 76; verse units, 249; weight and mass units, 184, 206; world temperature record, 259; world's most powerful light in, 250. *See also* Dunkirk; Paris; Troyes
Francis I (France): cloth measure, 29
freight: avoirdupois weight, 185
fruit: avoirdupois weight, 181; fundamental unit defined, xv

Gambia: United States customary system, xiv
Gaul: linear measure, 16; liquid measure, 93
Gauss, Karl Friedrich, 290
General Electric Company, 250–251

Germany: avoirdupois weight, 179, 185; count units, 224; light intensity unit, 252; linear measure, 17; metric length, 209, 210; printer's measure, 30; spirits measure, 199; metric weight and mass units, 219. *See also* Brunswick; East Germany; Hamburg; Prussia; Schleswig-Holstein; West Germany; Würtemberg

Ghana: United States customary system, xiv

Gilbert, William, 290

glass: avoirdupois weight, 180

gold: dry measure, 96; troy weight, 170; weight and mass units, 170, 201, 206

grain: avoirdupois weight, 176; capacity and volume units, 101, 149, 167; weight and mass units, 176, 206

gravimetry, 169

Gray, Louis H., 268

Great Britain: apothecaries' weight, 186; avoirdupois weight, 173–185; book size units, 228; capacity and volume units, 167, 168; cubic measure, 85, 87; count units, 223, 224; dry measure, 97, 98, 99, 101; earliest known unit of length, 1; Elizabethan and Middle English verse units, 247, 248, 249, 250; force units, 266, 267; linear measure, 5, 6, 7, 12, 13, 15, 57; liquid measure, 89–90, 91, 93; mariner's measure, 20; printer's measure, 5, 31, 32; spirits measure, 113, 116, 119, 121; surface units, 62, 63, 64, 65; troy weight, 170–173; world's most powerful searchlight, 250–251. *See also* British imperial system of measures; Greenwich; Ireland; Scotland; Stoke on Trent; Wales

Greece: avoirdupois weight, 173; capacity and volume units, 131–135, 167; linear measure, xiii, 5, 7, 9, 10, 35–40; metric length units, 209, 210, 211; metric surface units, 213; music units,

244; spirits measure, 116, 121; surface units, 75, 76; verse units, 246, 247, 248; weight and mass units, 173, 188–189, 193, 194, 204, 206. *See also* Attica

Greenwich (England), 232

Gregory XIII (Pope), 230

gross weight defined, xvii

Guatemala: avoirdupois weight, 182; capacity and volume units, 162; linear measure, 10, 49; weight and mass units, 206

gunpowder: avoirdupois weight, 184–185

Gunter, Edmund, 21

Gutenberg, Johan: printer's measure, 30

Haiti: linear measure, 9; spirits measure, 114

Hamburg, Germany, 19

Hawaii: calendars, 231

hay (old hay; new hay): avoirdupois weight, 177, 178, 183

heaped measure defined, xvii

Hebrew units. *See* bible: Israel (ancient)

Heisenbeg Werner: Uncertainty Priniple, xiii

Henry I (England): linear measure, 11

Henry III (England): troy weight, 170

Henry VIII (England): liquid measure, 88; square measure, 63

Henry, Joseph, 288

Hertz, Heinrich Rudolf, 281

Hindu states: binary system, xviii

Holland. *See* Netherlands

Honduras: linear measure, 10, 14, 57; surface units, 74, 76; weight and mass units, 206

Hong Kong: linear measure, 56

hops: avoirdupois weight, 181

horses: height of, 8; horse racing, 23

Hungary: capacity and volume units, 116, 166; linear measure, 7, 14, 57; spirits measure, 116; surface units, 64, 76

Iceland: capacity and volume units, 150; linear measure, 3, 5, 7, 14,

17, 55; mariner's measure, 21; surface units, 70–71; weight and mass units, 191, 206
imperial measure. *See* British imperial system of measures
India: avoirdupois weight, 182; capacity and volume units, 135–136, 167; cloth measure, 24; linear measure, 10, 11, 16, 40–41, 55, 56; weight and mass units, 182, 189, 205. *See also* Bengal; Calcutta; Ceylon; Hindu states; Madras; Rangoon; United Province
Indonesia: linear measure, 17. *See also* Java; Sumatra
International Astronomical Union, 17, 18
International Bureau of Weights and Measures, xv
International Electrical Congress (1893), 282, 283, 288
International Hydrographic Bureau, 15
International system of measures, xiii–xiv; cubic measure, 87
International System of Units, xiv
Iran: capacity and volume units, 167; metric weight and mass units, 219; weight and mass units, 205, 206, 219. *See also* Persia
Iraq: surface units, 76
Ireland: linear measure, 15, 56
iron: avoirdupois weight, 180
irrigation: cubic measure, 89
Israel (ancient): avoirdupois weight, 178; capacity and volume units, 113, 136–137; linear measure, 6, 10, 11; spirits measure, 113; surface units, 74; weight and mass units, 178, 206. *See also* bible
Italy: capacity and volume units, 168; linear measure, 14, 55; metric length units, 210, 211; surface units, 75; verse units, 249, 250; weight and mass units, 206; *See also* Naples; Rome (ancient); Trieste

Japan: avoirdupois weight, 174; capacity and volume units, 151;

linear measure, 9, 45–47; mariner's measure, 20, 21; surface units, 71–72, 77; weight and mass units, 174, 195–196
Java: avoirdupois weight, 174; linear measure, 14
Jeroboam I (king, ancient Israel): spirits measure, 113
Jersey (Channel Islands): linear measure, 52
jeweler: mass unit, 170
Joule, James Prescott, 263

Kassen Act, xiii
Kelvin, Lord, *see* Thomson, William
Kelvin temperature scale, 259
Kenya: linear measure, 11
Korea: printer's measure, 30

Lambert, Johann, 254
land: linear measure, 12; plowed furrow, 13, 23; square measure, 64; surface measure in South Africa, 181. *See also* earth (soil)
land appraisal: surveyor's measure, 21
lard: avoirdupois weight, 178
Latvia: capacity and volume units, 167; linear measure, 47, 48; surface units, 76, 77
leather: shoe leather, 4, 24
leather hides: avoirdupois weight, 185
length units: ampere-turn (electrical unit) per length units, 291–292; per capacity and volume units, 273–274; customary units, 1–57; earliest known unit of length, 1; early standards, xv; length defined, 1; length unit defined, 1; metric 208–211; volt (electrical unit) per length units, 291. *For customary units see also:* cloth measure units; linear measure units; mariner's measure units; printer's measure units; solar measure units; surveyor's measure units.
Libya: capacity and volume units, 151–152; surface units, 74, 76; weight and mass units, 205
light measure units, 250–254
light waves: linear measure, 2

Lilius, Aloysius, 230
lime: avoirdupois weight, 181; dry measure, 100
linear measure units, 2–18, 34–57; in Abyssinia, 56; in Aden, 11; in Arabia, 9, 13, 16, 34–35, 57; in Argentina, 5, 9, 10, 13, 14, 16, 53; in Assyria, 9, 10, 35, 42; in Austria, 5, 9, 14, 17, 57; in Bavaria, 5, 9, 57; in Belgium, 9, 12, 52; in Bohemia, 9; in Brazil, 7, 9, 10, 14, 16, 53; in Byelorussian SSR, 7, 47; in Calcutta, 8, 11, 55; in California, 10; in Canada, 7, 9; in Chaldea, 35, 42; in Chile, 5, 10, 16, 53; in China, 53–54; in Colombia, 10; in Cuba, 10, 16, 56; in Cyprus, 8; in Czechoslovakia, 57; in Denmark, 3, 5, 7, 9, 17, 54; in Dominican Republic, 57; in Dutch East Indies, 56, 57; in East Africa, 57; in Ecuador, 53; in Egypt, 10, 54, 57; in El Salvador, 9, 10; in Estonia, 7, 47, 48; in Finland, 7, 14, 17, 48; in France, 4, 5, 7, 9, 14, 16, 53, 56; in Gaul, 16; in Germany, 7, 17; in Great Britain, 5, 6, 7, 12, 13, 15, 57; in Greece, xiii, 5, 7, 9, 10, 35–40; in Guatemala, 10, 49; in Haiti, 9; in Honduras, 10, 14, 57; in Hong Kong, 56; in Hungary, 7, 14, 57; in Iceland, 3, 5, 7, 14, 17, 55; in India, 10, 11, 16, 40–41, 55, 56; in Indonesia, 17; in Ireland, 15, 56; in Israel (ancient), 6, 10, 11; in Italy, 14, 55; in Japan, 9, 45–47; in Java, 14; in Jersey (Channel Islands), 52; in Kenya, 11; in Latvia, 47, 48; in Louisiana, 10; in Malabar 56; in Malacca, 10, 56, 57; in Malta, 56; metric length units compared with, 211–212; in Mexico, 5, 7, 9, 11, 16, 49; in Mongolia, 12; in Naples, 14; in Netherlands, 7, 9, 14; in Nicaragua, 14; in Norway, 9, 56, 57; in Paraguay, 5, 9, 10, 16, 53, 56; in Paris, 52; in Persia, 9, 11, 16, 35, 40, 41–42, 57; in Peru, 10; in Philippines, 7; in Poland, 5, 9, 14, 55; in Portugal, 5, 7, 9, 14, 16, 53, 56; in Prussia, 3, 7, 9, 14, 17, 57; in Quebec, 12; in Rangoon, 56, 57; in Rome (ancient), 9, 10, 11, 13, 14, 42–45; in Rumania, 56, 57; in Russia, 5, 7, 9, 17, 47–48; in Scotland, 8, 14; in Siam, 5, 55–60; in Singapore, 10; in Somaliland, 10; in South Africa, 7; in Spain, 5, 7, 9, 10, 16, 48–50; in Sumatra, 14; in Sweden, 3, 5, 9, 17, 50–52; in Switzerland, 5, 7, 9, 12, 52–53; in Texas, 10; in Thailand, 57; in Turkey, 14; in United States, 11; in Uruguay, 10, 16, 53; in Venezuela, 14; in watchmaking, 56; in Yugoslavia, 9, 60
liquid measure units, 89–94; comparison with apothecaries' fluid, cooking, cubic, dry, and spirits measures (tables), 123–126 (size), 126–129 (intermeasure); metric units compared with, 217–218; origin of system, 90. *See also* apothecaries' fluid measure units; capacity and volume units; cooking measure units; spirits measure units
Lisbon (Portugal): capacity and volume units, 156
live load defined, xvii
long measure, 2
Louisiana: linear measure, 10
lumber: cubic measure, 82, 86, 87; weight and mass units, 205
Luxembourg: capacity and volume units, 167

Mach, Ernst, 271
Madras: capacity and volume units, 168; surface units, 75
Malabar: linear measure, 58
Malacca: linear measure, 10, 56, 57
Malaysia: avoirdupois weight, 174; liquid measure, 91; Malay verse form, 249
Malta: capacity and volume units, 167; linear measure, 56; weight and mass units, 193, 194, 206
mariner's measure units, 19–21; in Denmark, 20; in Finland, 21; in France, 20; in Great Britain, 20;

in Iceland, 21; in Japan, 20, 21; in
United States, 20
mass: defined, xvii, 169;
intermeasure comparison of
weight and mass units, 187–188;
jeweler's unit, 170; metric defined,
215; units of, 169–206
Massachusetts Institute of
Technology, 242
Maxwell, James Clerk, 289
measure: heaped (defined), xvii;
prototype standards, xv–xvi;
standard of (defined), xv; struck
(defined), xvii. *See* British
imperial system of measures;
metric system; United States
customary system of measures
Mechain, M., 210
mercury: avoirdupois weight, 178
Meter, Treaty of the, xiii
Metric Convention, xiii
Metric Conversion Act, xiv
metric system (*Le Système
International d'Unites*), 207–220;
capacity and volume units,
214–217; capacity and volume
units compared with customary
units (apothecaries' fluid, cubic,
dry, liquid, spirits measures),
217–218; flaws in, xii; fundamental quantitative units of, xiv;
length units, 208–211; length units
compared with linear measure
and solar measure, 211–212; mass
defined, 215; mass units compared
with customary weight and mass
units (apothecaries', avoirdupois,
troy weights), 219–220; prototype
standard of meter, xv; surface
units, 212–214; surface units
compared with customary surface
units, 214; in United States, xiv,
xvi; volume defined, 215; weight
and mass units, 218–219. *See also*
centimeter-gram-second
subsystem
Metric Usage in Federal
Government Programs, xiv
Mexico: avoirdupois weight, 179,
182; capacity and volume units,
152–153; linear measure, 5, 7, 9,
11, 16, 49; surface units, 65, 74,

76, 77; weight and mass units,
179, 182, 198, 199, 201, 202, 206
mining industry: square measure,
61; testing value of ore, 172
mks. *See* meter-kilogram-second
system
money: troy weight, 170
Mongolia: linear measure, 12
Mouton, Gabriel, xiii
music: Greek units, 244; units in,
243–245

nails: avoirdupois weight, 179
Naples: linear measure, 14
National Aeronautics and Space
Administration, 242
National Bureau of Standards
(Washington, D.C.), xiii
navigation: angular measure, 79
Nepal: surface units, 77
net weight defined, xvii
Netherlands: avoirdupois weight,
173; capacity and volume units,
153–154; dry measure, 101; linear
measure, 7, 9, 14; metric capacity
and volume units, 216; metric
length units, 209, 210, 211; metric
surface units, 213; spirits
measure, 115, 116, 119; metric
weight and mass units, 218;
surface units, 75, 76, 213; weight
and mass units, 173, 196–197, 218
Newton, Sir Isaac, 266
Nicaragua: avoirdupois weight, 182;
capacity and volume units, 167;
linear measure, 14; surface units,
73, 74, 76, 77; weight and mass
units, 182, 206
North Borneo. *See* Sabah
Norway: avoirdupois weight, 179;
capacity and volume units, 145,
146, 168; linear measure, 9, 56,
57; square measure, 62; weight
and mass units, 179, 191, 206
nuclear physics: square measure,
59

oatmeal: avoirdupois weight, 180
offshore territorial limit
determination of, 16
Ohm, Georg Simon, 285
Ohm's law, 283

oil, vegetable: capacity and volume units, 160
Oman: United States customary system, xiv
Omnibus Trade and Competitiveness Act, xiv
Ondrejka, Rudolf, 225
ore: avoirdupois weight, 183; testing value of, 172

Palestine. *See* Israel (ancient)
Panama: capacity and volume units, 161
paper measure, 226–227
Paraguay: capacity and volume units, 93, 167; linear measure, 5, 9, 10, 16, 53, 56; surface units, 75
Paris: linear measure, 52
Paris Congress (1894), 285
Persia: capacity and volume units, 130, 131, 167; linear measure, 9, 11, 16, 30, 40, 41–42, 57; metric length units, 209, 211; metric surface units, 213; surface units, 76; weight and mass units, 206. *See also* Iran
Peru: avoirdupois weight, 179, 182; linear measure, 10; surface units, 73, 75, 77; weight and mass units, 179, 182, 199, 202
petroleum: cubic measure, 83; liquid measure, 93
Philippines: capacity and volume units, 166; linear measure, 7; surface units, 75, 76; weight and mass units, 200, 205
photometry, 250
physics. *See* nuclear physics
Pi Sheng, 30
planimetry, 77
plowing: square measure, 63
Poland: capacity and volume units, 155; linear measure, 5, 9, 14, 55; surface units, 76, 77; weight and mass units, 197–198, 199, 200
pork: avoirdupois weight, 181
Portugal: capacity and volume units, 155–157; linear measure, 5, 7, 9, 14, 16, 53, 56; metric length units, 210; surface units, 76; weight and mass units, 198–199, 202. *See also* Lisbon

prefixes of units (metric and miscellaneous), 208
pressure units, 256–258
printer's measure units, 30–34; in China, 30; in Germany, 30; in Great Britain, 5, 31, 32; in Korea, 30
Prussia: dry measure, 101; linear measure, 3, 7, 9, 14, 17, 57; square measure, 63
Puerto Rico: surface units, 74, 75

Quebec: linear measure, 12

radiant power unit, 272
radiation units, 267–268
Rangoon: linear measure, 56, 57
Rankine, William, 259
Rankine temperature scale, 259
reactive power unit, 272
de Reaumur, R. A. F., 259
Reaumur (Reaumur, Reaum) temperature scale, 259
rice: avoirdupois weight, 174
Richard I (England): length and capacity standards, xv
Roentgen, Wilhem Konrad, 267
Rome (ancient): avoirdupois weight, 173, 174; calendars, 230; capacity and volume units, 116, 134, 137–140; duodecimal system, xviii; Latin verse units, 247, 248; linear measure, 9, 10, 11, 13, 14, 42–45; spirits measure, 116; surface units, 63, 67–69; weight and mass units, 173, 174, 189–190
roofing: square measure, 61
Rumania: capacity and volume units, 167; linear measure, 56, 57
Russia: apothecaries' weight, 186; avoirdupois weight, 177; capacity and volume units, 112, 157–160; laboratory-pressure record, 256; linear measure, 5, 7, 9, 17, 45–47; spirits meaure, 112; surface units, 76; weight and mass units, 177, 186, 198, 199–200; world temperature record, 259; world's longest-burning lamp, 250

Sabah (North Borneo): liquid measure, 92

salt: avoirdupois weight, 181; cooking measure, 107

Schleswig-Holstein: surface units, 77

Scotland: avoirdupois weight, 180; dry measure, 96, 99; linear measure, 8, 14; small liquor bottles, 110; square measure, 62, 63

sexagesimal system, xvi, xviii

shipping: cubic measure, 83, 87

shipping, marine: cubic measure, 85

shoe leather: cloth measure, 24; linear measure, 4

Siam: avoirdupois weight, 174; linear measure, 5, 55–60; metric length units, 210; surface units, 62, 76; weight and mass units, 174, 206. *See also* Thailand

Siegbohn, Georg, 2

von Siemens, Alfred Ernst Werner, 273

Sierra Leone: avoirdupois weight, 183; United States customary system, xiv

silver: weight and mass units, 201, 206

Singapore: linear measure, 10

Skewes, Stanley, 225

solar (stellar) measure units, 18–19; metric length units compared with, 211–212

Somaliland: capacity and volume units, 167; linear measure, 10; surface units, 76

sound measure units, 241–243

sound velocity, 241

South Africa: capacity and volume units, 165, 168; linear measure, 7; surface measure, 63, 181; weight and mass units, 205

Soviet Union, *See* Russia

Spain: avoirdupois weight, 177; capacity and volume units, 115, 160–163; linear measure, 5, 7, 9, 10, 16, 48–50; metric length units, 210; spirits measure, 115; surface units, 72–74; weight and mass units, 177, 199, 200–202

speed units, 270–271. *See also* velocity units

spirits measure units, 110–122;

comparison with apothecaries' fluid, cooking, cubic, dry, and liquid measures (tables), 123–126 (size), 126–129 (intermeasure); metric units compared with, 217–218; small liquor bottles marketed, 110. *See also* capacity and volume units

square measure units, 59–65; comparative areas of square measure units (table), 65–66; metric surface units compared with customary units, 214. *See also geographical references under* surface units

steam: atmospheric pressure and, 258

steel: avoirdupois weight, 180

stellar measure. *See* solar measure

Stoke on Trent (England): oversize sherry bottle, 114

stone: cubic measure, 84

straw: avoirdupois weight, 177, 183

streams: water-flow units, 270

struck measure defined, xvii

subatomic measure, 2

Sumatra: avoirdupois weight, 174; linear measure, 14

surface units, 58–80; in Annam, 76, 77; in Arabia, 75; in Argentina, 75, 213; in Austria, 64, 76; in Bavaria, 63; in Bengal, 75; in Bohemia, 76; in Brazil, 75; in Bulgaria, 76; in Canary Islands, 73; in Chile, 75; in China, 76, 213; in Costa Rica, 74, 76; in Cuba, 74, 75, 77; in Cyprus, 74; in Czechoslovakia, 64, 76; in Denmark, 69–70; in Dominican Republic, 77; in Dutch East Indies, 75; in Egypt, 75; in El Salvador, 74; in Estonia, 76, 77; in Finland, 77; in France, 63, 76; in Great Britain, 62, 63, 64, 65; in Greece, 75, 76, 213; in Honduras, 74, 76; in Hungary, 64, 76; in Iceland, 70–71; in Iraq, 76; in Israel (ancient), 74; in Italy, 75; in Japan, 71–72, 77; in Latvia, 76, 77; in Libya, 74, 76; in Madras, 75; metric, 212–214; metric compared with customary, 214; in Mexico, 65, 74, 76, 77; in Nepal,

77; in Netherlands, 75, 76, 213; in Nicaragua, 73, 74, 76, 77; in Norway, 62; in Paraguay, 75; in Persia, 76, 213; in Peru, 73, 75, 77; in Philippines, 75, 76; in Poland, 76, 77; in Portugal, 76; in Prussia, 63; in Puerto Rico, 74, 75; in Rome (ancient), 63, 67–69; in Russia, 76; in Schleswig-Holstein, 77; in Scotland, 62, 63; in Siam, 62, 76; in Somaliland, 76; in South Africa, 63, 181; in Spain, 72–74; surface measures defined, 58; in Sweden, 76; in Switzerland, 63; in Texas, 64, 65; in Turkey, 74; in United Province (India), 75; in United States, 61, 65; in Uruguay, 75, 77; in Venezuela, 73; in Wales, 75; in West Germany, 77; in Würtemberg, 63; in Yugoslavia, 64, 74. *See also* angular measure; square measure units

Surinam: cloth measure, 25

surveying: square measure, 62, 63; surface unit in South Africa, 181

Surveyor's Act, xiii

surveyor's measure units, 21–23

Sweden: avoirdupois weight, 179; capacity and volume units, 163–165; linear measure, 3, 5, 9, 17, 50–52; metric length units, 211; surface units, 76; weight and mass units, 179, 202–203

Switzerland: avoirdupois weight, 179, 180; capacity and volume units, 165–166; cloth measure, 25; linear measure, 5, 7, 9, 12, 52–53; metric length units, 209; metric weight, 219; square measure, 63

Syria: capacity and volume units, 167

Systems of measure in general use. *See* British imperial system of measures; metric system; United States customary system of measures

tea: avoirdupois weight, 174

temperature: Celsius (centigrade) scale, xvii, 259; Fahrenheit scale, xvii, 259; Kelvin scale, 259; Reaumur scale, 259; units of, 259–261

Tesla, Nikola, 290

Texas: linear measure, 10; square measure, 64, 65

Thailand: avoirdupois weight, 174; capacity and volume units, 168; linear measure, 57; weight and mass units, 174, 205. *See also* Siam

Thomson, William (Lord Kelvin), 259, 286

time measure: atomic hydrogen master (timekeeper), 233; basis of, 228; calendars, 229–231; decimal time, xvi, 231; standard time (U.S.), 232; sun time, 232; time zones, 231–232; units of, 233–241; universal coordinate time, 229; variations affecting, 228–229

TNT: avoirdupois weight, 184

tobacco: weight and mass units, 205

Tonga: United States customary system, xiv

Trieste: spirits measure, 116

Trinidad: United States customary system, xv

troy weight units, xiii, 170–173; comparison with apothecaries' and avoirdupois units, 187–188; metric mass units compared with, 219–220

Troyes (France), 170

Tunisia: weight and mass units, 193, 194

Turkey: capacity and volume units, 167; linear measure, 14; metric length units, 210; surface units, 74; weight and mass units, 192, 193, 194, 205

typesetting. *See* printer's measure units

Uncertainty Principle, xiii

unit, derived: defined, xv

unit, fundamental: defined, xv

unit of measurement: defined, xv

unit prefixes (metric and miscellaneous), 208

United Kingdom. *See* Great Britain

United Province (India): surface units, 75

United States: avoirdupois weight, 174, 179, 181, 182; book size

units, 228; count units, 223, 224;
cubic measure, 86; dry measure,
101; linear measure, 11; mariner's
measure, 20; metric system, xiv,
xvi, 220; metric weight (U.S. Post
Office), 220; momentary-pressure
record, 256; offshore territorial
boundary determination, 16;
rotary speed record, 280; square
measure, 61, 65; time zones, 232.
See also California; Cambridge;
Hawaii; Louisiana; Texas; United
States customary system of
measures; Washington, D.C.
United States customary system of
measures: vs. British imperial
system, xv; fundamental
quantitative units of, xiv
U.S. Naval Research Laboratory
(Washington, D.C.), 233
U.S. Type Founders Association, 4
Uruguay: capacity and volume
units, 166; linear measure, 10, 16,
53; surface units, 75, 77
USSR. *See* Russia

vegetables: avoirdupois weight, 181
velocity units, 270–271
Venezuela: avoirdupois weight, 182;
capacity and volume units, 162,
166–167; linear measure, 14;
surface units, 73; weight and
mass units, 182, 199, 202
verse: English, 247, 248, 249, 250;
feminine rhyme, 247; French, 249;
Greek, 246, 247, 248; Italian, 249,
250; Latin, 247, 248; Malay, 249;
units of, 246–250
vibrational energy unit, 271–272
Violle, Jules, 252
Virginia, University of, 280
viscosity units, 268
Volta, Alessandro, 282
volume: comparison of apothecaries'
fluid, cooking, cubic, dry, liquid, and
spirits measures (tables), 123–126
(size), 126–129 (intermeasure);
defined, xvii, 82; effect of
temperature and gravity on, 82;
metric unit defined, 215; mho
(electrical unit) per volume units,
293–294; ohm (electrical unit) per

volume units, 293; units of, 81–168.
See also capacity and volume units

Wales: capacity and volume units,
168; surface units, 75
wallpaper: cloth measure, 27;
square measure, 61
Wang Chen, 30
Washington, D.C., 233
watchcases: troy weight, 170
watchmaking: linear measure, 56
water: mass/heat unit, 272
water-flow units, 269–270
Watt, James, 277, 279
Weber, Wilhelm Eduard, 289
weight: dead load (defined), xvii;
defined, 169; earliest known
weight standard, 170; gross
weight (defined), xvii; live load
(defined), xvii; net weight
(defined), xvii; units of, 169–206.
See also apothecaries' weight;
avoirdupois weight; troy weight
units; weight and mass units
weight and mass units: in Algeria,
206; in Angola, 179; in Arabia,
206; per area units, 274; in
Argentina, 179, 198, 200, 202,
206; in Attica (Greece), 178; in
Austria, 180, 204, 219; in
Babylonia, 178; in Bavaria, 204,
206; in Bolivia, 177, 199, 202; in
Brazil, 179, 182, 198, 199, 200,
202, 206; in Brunswick
(Germany), 179–180; in Bulgaria,
194, 204; capacity and volume
units per weight and mass units,
273; in Chile, 179, 199, 200, 202;
in China, 174, 206; in Colombia,
179, 182, 198, 200, 202, 204; in
Costa Rica, 182, 206; in Cuba,
199, 202; in Cyprus, 193, 194, 206;
in Denmark, 179, 180, 191–192,
202, 203; in Ecuador, 199, 202; in
Egypt, 170, 182, 192–194; in El
Salvador, 179, 206; in Finland,
203; in France, 184, 206; in
Germany, 179, 185, 219; in Great
Britain, 170–186 *passim*, 220; in
Greece, 173, 188–189, 193, 194,
204, 206; in Guatemala, 182, 206;
in Honduras, 206; in Iceland, 191,

206; in India, 182, 189, 205;
intermeasure comparison of
weight and mass units, 187–188;
in Iran, 205, 206, 219; in Israel
(ancient), 178, 206; in Italy, 206;
in Japan, 174, 195–196; in Java,
174; in Libya, 205; in Malaysia,
174; in Malta, 193, 194, 206;
metric, 218–219; in Mexico, 179,
182, 198, 199, 201, 202, 206; in
Netherlands, 173, 196–197, 218;
in Nicaragua, 182, 206; in
Norway, 179, 191, 206; in Persia,
206; in Peru, 179, 182, 199, 202;
in Philippines, 200, 205; in
Poland, 197–198, 199, 200; in
Portugal, 198–199, 202; in Rome
(ancient), 173, 174, 189–190; in
Russia, 177, 186, 198, 199–200; in
Scotland, 180; in Siam, 174, 206;
in Sierra Leone, 183; in South
Africa, 205; in Spain, 177, 199,
200–202; in Sumatra, 174; in
Sweden, 179, 202–203; in
Switzerland, 179, 180, 219; in
Thailand, 174, 205; in Tunisia,
193, 194; in Turkey, 192, 193, 194,
205; in United States, 174, 179,
181, 182, 220; in Venezuela, 182,

199, 202; in Yugoslavia, 194,
203–204
Weights and Measures, General
Conference on, xiv, 233
West Germany: capacity and volume
units, 167; linear measure, 7;
surface units, 77
wheat: dry measure, 99, 100
whisky: spirits measure, 110, 111
Winchester bushel, xiii
wine: Bordeaux, 114; capacity and
volume units, 160; liquid
measure, 90; port, 114; sherry,
114; spirits measure, 112, 114,
115, 116, 117, 118, 119
wire: cubic measure (electrical unit),
82; linear measure, 4
wool: avoirdupois weight, 175, 176,
177, 181, 182, 184, 185; cloth
measure, 27–28, 28–29, 30
Würtemberg (Germany): square
measure, 63

Yemen, Southern: United States
customary system, xiv
Yugoslavia: capacity and volume
units, 167; linear measure, 9, 60;
surface units, 64, 74; weight and
mass units, 194, 203–204